ptimization techniques for electrical machines

Dieter Gerber

List of Figures

List of Tables

Abbreviations

ANN	Artificial Neural Network
ANOVA	Analysis of Variance
BLUE	Best Linear Unbiased Estimation
CAE	Computer-Aided Engineering
CI	Chebyshev Interval
CPM	Claw Pole Motor
DACE	Design and Analysis of Computer Experiments
DEA	Differential Evolution Algorithm
DFSS	Design for Six Sigma
DG	Differential Gear
DPMO	Defects Per Million Opportunities
DTC	Direct Torque Control
EA	Evolutionary Algorithms
EDAs	Estimation of Distribution Algorithm
EMF	Electromotive Force
EV	Electric Vehicle
FOC	Field-Oriented Control
GA	Genetic Algorithm
HEV	Hybrid Electric Vehicle
HM	Homogenization Method
ICE	Internal Combustion Engine
IPMSM	Interior Permanent Magnet Synchronous Motor
LSL	Lower Specification Limit
LSM	Least Square Method
MCA	Monte Carlo Approach
MDO	Multi-disciplinary Design Optimization
MLE	Maximum Likelihood Estimation
MMA	Method of Moving Asymptotes
MPC	Model Predictive Control

MPP	Most Probable Point
MTPA	Maximum Torque Per Ampere
NSGA	Non-dominated Sorting Genetic Algorithm
OC	Optimality Criteria
PCCI	Polynomial Chaos Chebyshev Interval
PCE	Polynomial Chaos Expansion
PID	Proportion Integration Differentiation
PM	Permanent Magnet
PMSM	Permanent Magnet Synchronous Motor
POF	Probability of Failure
PSO	Particle Swarm Optimization
RBDO	Reliability-Based Design Optimization
RBF	Radial Basis Function
RDO	Robust Design Optimization
RSM	Response Surface Model
SA	Sensitivity Analysis
SIMP	Simplified Isotropic Material with Penalization
SM	Space Mapping
SMCA	Scan and Monte Carlo Approach
SOM	Sequential Optimization Method
SPMSM	Surface-mounted Permanent Magnet Synchronous Motor
SS	Switching Signal
SVM	Support Vector Machine
TFM	Transverse Flux Motor
USL	Upper Specification Limit

Abstract

To investigate the design optimization techniques of electrical machines, a literature survey is conducted about the problem modeling, and techniques utilized for the effective optimization conduction with various case studies. As a comprehensive design optimization example, an in-wheel motor development for distributed direct vehicle driving is investigated in detail. The works on the application analysis, new material application (grain-oriented silicon steel), topology development, manufacturing process, experiment verification, and parametric deterministic optimization are presented.

Based on the state-of-art design optimization methods and case studies, challenges and proposals are also presented in the survey. In the design stage, the bring-up of new topology depends on the expertise of the designers. This means the expert system still plays an important role in the application-oriented design optimization process. Moreover, parametric optimization is carried out based on the specific design which means the freedom of optimization is limited. To overcome the restriction of parametric optimization for high freedom optimization topology optimization method is investigated. At the current stage, the topology optimization of the soft magnetic components of electrical machines considering various electromagnetic performances is studied. The optimization results of the design examples verified the effectiveness of the proposed method in achieving the optimal shape of the design.

Another problem is about the uncertainties in manufacturing such as tolerances which bring in reliability problems for the conventional deterministic optimal solution. Under this circumstance, the robustness optimization is important for searching optimal solution with both high objective performance and reliability. For the robust optimization of electrical machines, the additional uncertainty quantification and robustness assessment further aggravating the complexity and computation cost of the problem. Considering the problem with different types of uncertainties, high effective optimizers are proposed based on effective uncertainty quantification methods with a general framework. The numerical study results proved the effectiveness of the proposed method.

Contents

Chapter 1 Introduction

1.1 Background and Significance

Electrical machines are the electromechanical energy converter coupled with the magnetic field. As the hearts of various power generation and drive systems, the development of electrical machines has a history of over a century. As presented in Figure 1-1, electrical machines consume more than 40 percent of the total electricity generated globally [1-3]. Compared with other applications, the electricity consumption of electrical machines occupies the largest proportion. From the aspect of energy consumption and environmental protection, the efficiency of the electrical machines is a crucial point for the sustainable development of humanity. In addition to the high efficiency, the performance of high torque/power density, low torque ripple, low cost, etc. are generally pursued by the designers while ensuring to meet the specifications of the application[4-6]. To find the optimal design, the development of the design and optimization techniques of electrical machines has always been the hotspot for the industry and academics.Equation Chapter 1 Section 1

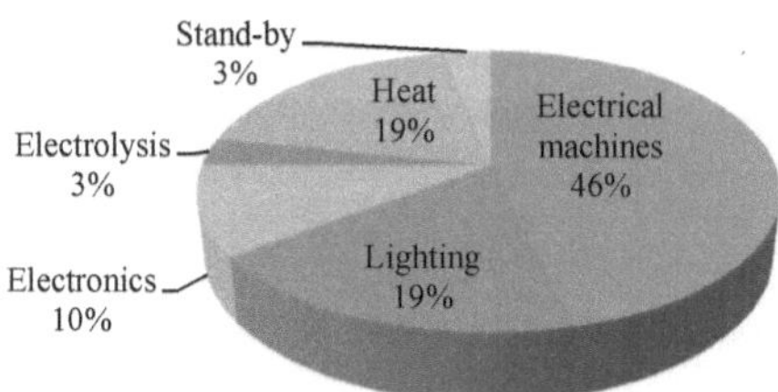

Figure 1-1 Global electricity demand by sector and end-use.

Electrical machines are generally designed and optimized to meet some specific application demands required by the application. After the determination of the requirements, the design optimization of electrical machines starts with proposing or selection of electrical machine type, material, and structures. This application-oriented design process determines whether the final design is successful. On the other hand, the exploration of the novel topology, operation rules, and new materials will enrich the selection spaces for different applications.

Based on the initial design, the following optimization processes are to find the optimal design considering the specifications. Conventional parametric optimization based on the structure of the initial design, such as the size optimization and shape optimization, is generally with low freedom. That means the limitation of the optimization process on the performance enhancement and the dependence on the initial design. Compared with the parametric optimization methods, topology optimization shows the highest optimization freedom for obtaining the optimal design with a new shape. However, as a new kind of optimization method, the research on the topology optimization for electrical machines is still at an early stage. Effective topology optimization methods should be developed for electrical machines.

At the end of the electrical machine development procedure, the optimal design is used for production. However, the inevitable uncertainties such as the tolerances in the manufacturing, material diversities, and variation of the working environment, result in the performance variation between the deterministic optimal design and the prototype. Without considering the uncertainties, the deterministic optimal solution with high sensitivity to the uncertainties leads to the high diversity of the performance or failures because of the violation of the constraints due to the performance variation. Taking full consideration of the uncertainties, robust optimization method development is important for shortening the development period of the electrical machines and manufacturing the product with high quality. Nevertheless, due to the additional robustness assessment in the optimization process, the computation burden of the deterministic optimization problem caused by multidisciplinary modeling, high dimension, and nonlinear calculation, is aggravated further. Therefore, exploring effective robust optimization methods for electrical machines is also one of the goals of this research.

1.2 Research Objectives

According to the brief introduction above, the specific research objectives are given as follows:

(1) To conduct a comprehensive literature survey on the design optimization techniques of electrical machines with case studies.

(2) To explore the application of new core material (grain-oriented silicon steel) in the design optimization of a dual rotor axial flux electrical machine for the in-wheel motor drive.

(3) To investigate the topology optimization method for electrical machines with a high degree of design optimization freedom.

(4) To develop efficient general optimization methods considering the uncertainties in manufacturing for the robust optimization of electrical machines.

1.3 Outline of the Book

The following chapters are organized as follows

Chapter 2 presents a literature survey on the design optimization of electrical machines. The basic procedure, modeling, optimization techniques are reviewed at first with case studies. According to the development of the state-of-art techniques, the challenges and proposals are presented subsequentially.

Chapter 3 proposes a case study on an in-wheel motor development with the new core material, i.e. grain-oriented silicon steel. The design analysis process for the dual rotor axial flux motor with grain-oriented material is introduced in detail.

Chapter 4 presents the manufacturing methods and processes of the designed motor. The prototype test is conducted for verification of the simulation results. Based on the model and experiment results, the optimization is conducted for improving the performances.

In chapter 5, the topology optimization based on density method or so-called Simplified Isotropic Material with Penalization (SIMP) method is investigated for the design optimization of electrical machines with a high degree of freedom. The topology optimization of the ferromagnetic components in permanent magnet (PM) synchronous machines is currently researched which includes the sensitivity analysis of the electromagnetic performances with respect to the design variables and the establishment of the optimization framework. Two design examples are optimized, and the results validate the feasibility of the proposed optimization method.

Chapter 6 proposes the general robust optimizer based on intelligent algorithms and effective uncertainty quantification is established for effective robust optimization of electrical machines. For different kinds of uncertainties, including the stochastic, interval, and hybrid uncertainties, various uncertainty quantifications are introduced. They are Polynomial Chaos Expansion (PCE) and dimension reduction method for stochastic uncertainties, Chebyshev Interval (CI) method for interval uncertainties, and Polynomial Chaos Chebyshev Interval (PCCI) method for the hybrid uncertainties.

Chapter 7 concludes the work of the Book and summarizes the future works.

Chapter 2 Literature Survey on the Design Optimization of Electrical Machines

2.1 Introduction

This chapter aims to present a literature survey on state-of-art design optimization techniques and a discussion on the challenges and future directions of the design optimization methods for electrical machines. Section 2.2 introduces the basic design optimization procedure and the following sections 2.3-2.6 review the determination of initial design, multidisciplinary modeling and performance calculation for the constraints and objectives of the optimization model, selection of design variable and optimization algorithms respectively. Due to the heavy computation burden in the optimization caused by the high dimension, performance calculation and iteration, the techniques and strategies for the optimization acceleration are reviewed in section 2.7. Section 2.8 presents the rethinking, challenges, and proposals from the author and the research team. The following sections 2.9-2.11 introduce them in detail which are about topology optimization, robust optimization and application-oriented system-level optimization for electrical machines. The conclusions are given in section 2.12.

2.2 Basic Procedure for Design Optimization of Electrical Machines

The general process of design optimization of electrical machines includes two main stages, design and optimization. The main aim of the design stage is to find the feasible scheme (or several schemes) and design for a given application through the expert knowledge and investigation of different materials and dimensions, motor types and topologies, and multi-disciplinary analysis, etc. The analysis of this stage will provide information on the performances of the proposed design for the development of optimization models that will be used in the optimization stage.

The main target of the optimization stage is to improve the performance of the motor proposed in the design stage through optimization algorithms and methods. As the outcome, an optimal solution will be obtained for the single objective design situation, and some non-dominated solutions (called Pareto optimal solutions) will be gained for a multi-objective design situation after the completion of this stage. There is no fixed procedure for the design optimization of electrical machines. However, there are some common steps to be followed. These steps are briefly described as follows which may also interact with each other.

Step 1: Determine the specifications from the application. The performance requirement of the application may include the speed range, output torque/power, and working condition, etc. These specifications will provide the instructions for the design scheme selection.

Step 2: Select/determine possible initial designs including motor types, topologies, and materials qualitatively according to the specifications of the application. The main aim of this step is to obtain various motor options that may be suitable for a specific application. This process is also the dimension reduction process based on the expert system for the following optimization. Because the initial design is given by going through the possible designs for an application quantitatively, the computation burden is usually heavy. Therefore, there is a high requirement for the designer experience and the expert system on determining the limited initial designs effectively according to the application requirement.

Step 3: Implement multidisciplinary modeling and analysis for the potential designs and evaluating the performance included in the objectives and constraints. This process is coupled with the initial design proposing. Due to the multidisciplinary property, the performances in terms of different disciplines have to be investigated in this step which mainly includes electromagnetics, thermotics [1 2], reliability[3, 4], economics[4], mechanics [1, 5], dynamics[6,7], and acoustics[7,8].

Step 4: Model the optimization problem according to the design specifications, initial designs, and multidisciplinary performance modeling. Quantitively define the objectives and constraints and select the design variables.

Step 5: Conduct optimization and obtain an optimal solution or some optimal solutions. The implementation covers optimization algorithms and methods. For the optimization algorithm, the global optimum searching ability and convergence speed are generally regarded as the evaluation and selection criterion. Due to the complexity of the optimization modeling and computation burden, optimization strategies and methods are also required as effective complements for the algorithm. After the optimization, compare all designs and output the best one with detailed design parameters and the simulated performances.

Step 6: Validate the design with prototypes and experiment results. The results will provide the verification of the optimization and basis for the model modification for the next loop of optimization containing the above steps until determining the final optimal design.

We conclude the design optimization into the mathematical model which can be expressed as

$$\begin{aligned} \min: \quad & \mathbf{f}(\mathbf{x}) \\ \text{s.t.} \quad & g_j(\mathbf{x}) \le 0, j = 1, 2, \ldots, m \\ & \mathbf{x}_l \le \mathbf{x} \le \mathbf{x}_u \end{aligned} \tag{2-1}$$

where $\mathbf{x}, \mathbf{f}$ and g_j are the design parameter vector, objectives and constraints, and $\mathbf{x}_l$ and $\mathbf{x}_u$ are the lower and upper boundaries of the design variables, respectively. As written with different colors, we separate the optimization problem into four main steps, i.e. determination of the initial design, definition, and calculation of objective and constraints, selection of design variables, optimization by algorithms. The following introduction is about the problems in this procedure.

2.3 Determination of the Initial Design

The initial design determination of the electrical machine is based on the specifications extracted from the applications. Considering the specifications, the specific initial designs are then proposed considering machine types, structures, and material, etc.

2.3.1 Determination of Specifications from the Applications

Different applications may have exclusive requirements. The general requirements include the working speed and output torque, etc. Taking the electrical vehicle drive as an example, the speed range of the vehicle is roughly 0-200km/h. By applying the in-wheel drive, the operation speed of the in-wheel motor is estimated as 0-1800 rpm. However, if a centrally arranged electric motor is adopted with differential gears, the operation range of the motor and design space will be different from the in-wheel motor while the wheel speed should still be promised. The determination of the specifications provides guidance for the design scheme selection with different initial designs.

2.3.2 Determination of Machine Types, Materials, and Structures

As mentioned above, after determining the specifications of the application, the design scheme can be settled. The design falls into the specific electrical machine selection and design process. There is no unified method for the classification of electrical machines. According to the excitation method, there are direct current (DC) and alternating current (AC) machines; According to the motion type, there are rotating and linear machines; According to the operation principle, there are synchronous and asynchronous machines. For the electromagnetic design of electrical machines, the commonly used materials include conductor materials, soft magnetic materials, and PMs, while each type of material is differentiated into various categories based on the properties. Consequently, there are also machines named after the material such as the superconducting machines and PM machines, etc. Based on the structure or topology, the categories of electrical machines vary a lot. There are radial flux machines, axial flux machines, and transverse flux machines according to the flux path. In terms of the relative position of the stator and rotor, there are inner and outer rotor machines. According to the number of the stator and rotor, there are the single and double stator or rotor machines, etc.

Even though there is no uniform classification method, we can summarize here that the classification of the electrical machines is generally based on the material, excitation, and structure. The selection of the electrical machines for specific application is depended on the specific requirements. Generally, the PM machines are with higher power/torque density while the material cost is high. There is also the risk of demagnetization which leads to the reliability problems. Compared with PM machines, induction machines are of lower power density and cost yet higher reliability. The determination of the initial design requires the choice of one or several types of machines followed by the quantitative modeling and analysis of the electrical machine such as the combination of slots and poles, dimension with the performance modeling techniques in the following section until the initial designs are decided.

2.4 Multidisciplinary Modeling for Objective and Constraint Performances

Electrical machines are the electromechanical energy converters that transfer the power between electricity and mechanical power by the magnetic field. The principle of operation of electrical machines is based on electromagnetic theory. Therefore, the electromagnetic design is critical. The objectives and constraints usually contain electromagnetic performances including efficiency, cogging torque, torque ripple, power factor, and voltage with other general considerations about cost, mass, volume, etc.[9, 10].

The design of electrical machines is also a multi-physics problem. The design and optimization aim is to find the optimal solution considering the electromagnetic characteristics, temperature distribution, structural stress, vibration, noise, etc.[3-6, 11].

For the thermal design, the main aim is to calculate the temperature distribution in the machine according to the loss distribution and heat dissipation condition. Structure design aims to consider the stress and deformation of the machine under the electromagnetic and thermal analyses which leads to the vibration and noise problem [7, 8]. On one hand, the multi-physics modeling and analysis are of vital importance for pursuing the optimal solution and avoid the failure design or the design with high redundancy and for shortening the electrical machine development time. The multi-physics performance can be particularly important for special electrical machines. For example, in the high-speed motor design, the problems in the rotor stress, dynamics, cooling, and noise, etc. are as important as the electromagnetic analysis for pursuing the optimal solution. On the other hand, suitable thermal and mechanical structure design such as cooling structure design will enhance the electromagnetic performances and increase the power density.

Effective performance modeling methods for multi-physics performance modeling are required for the design and optimization process. To obtain the electromagnetic analysis, the analytical model and magnetic circuit method provide a quick performance calculation [1] [12-14]. However, due to the limitation of the accuracy, model establishing cost, and performance estimation ability, the generality of these methods is not high. Nevertheless, the well-established analytical model and magnetic circuit model can provide effective approaches for reducing the design optimization space. The circuit model is also suitable for the performance calculation of the field, such as the thermal network model for temperature field calculation which can be applied for the constrained thermal performance calculation [3, 15, 16].

With the development of computer-aided engineering (CAE) software, the finite element model (FEM) has been widely employed as a powerful tool for the multi-physics design and analysis of electrical machines. It can be used to analyze the coupled field in machines, such as electromagnetic-thermal, electromagnetic structure and thermal-structure[17, 18]. However, even with high accuracy, FEM is time-consuming for the design model of the high scale of elements. The multi-physics analysis is usually conducted by the iteration between different field computing results until the convergence which aggravates the computational cost further. A practical model reduction method is to replace the three-dimensional FEM with the two-dimensional model. Moreover, conducting multi-physics optimization generally requires multiple effective techniques such as parallel computing.

2.5 Selection of Design Optimization Variables

Parametric optimization is widely adopted for the design optimization of electrical machines. After the topology of the initial design is proposed, the geometry is parameterized at first. Theoretically, all material parameters and dimensions used in the design stage can be optimized as the design variables, such as remanence, length, width and height of PMs, radiuses of rotor and stator, number of turns of winding and the length of the air gap, etc. optimization of the electrical machines is to find the optimal combination of the design parameters.

Even though this design variable definition method is generally used, the drawback of the parametric optimization is obvious. On one hand, to describe a design, there are usually tens of parameters required. The sampling number of this high dimension design problem is huge which leads to a high computation burden. On the other hand, the optimization of freedom is limited. Once the geometry of the initial design is parametrized, the parametric optimization can hardly produce new topology or shape. For example, holes cannot arise in the rotor if no relevant geometry parameters are defined. Therefore, the experience of the designer or expert knowledge is also important in this case. This problem can be made up to some extent by defining the geometry parameters in more detail. However, the design parameter may increase further which leads to the computational burden in the optimization process.

Compared with parametric optimization, topology optimization is regarded as the non-parametric optimization method which aims to optimize the material layout of the initial design. The topology optimization based on FEM takes the material property of each element as the design variable for higher optimization freedom. The thoughts about the topology are discussed in section 2.9.

2.6 Optimization Algorithms

2.6.1 Popular Algorithms

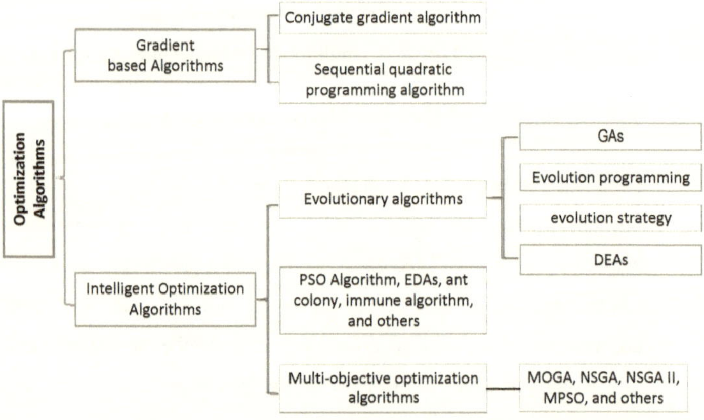

Figure 2-1 Popular optimization algorithms for the optimization of electrical machines.

The optimization algorithm is the program conducted iteratively for the optimal solution with the best objective performance subject to the constraints. The criteria to judge the effectiveness of the optimization algorithms includes the convergence speed and global convergence ability. Figure 2-1 shows some popular optimization algorithms for the design optimization of electrical machines as well as other electromagnetic devices. As shown, there are two main types, i.e. gradient-based algorithms and intelligent optimization algorithms.

As a kind of deterministic algorithm, the iteration of the gradient-based algorithm is driven by the gradient information obtained from the objective functions and constraints by the analytical calculation or the difference method. The application can be found in some works with motor performances estimated based on the magnetic circuit model [19-21]. Generally, deterministic algorithms can quickly convergence to the local extreme point in a small number of iteration steps. The ability to search for the global optimal solution of the deterministic algorithms cannot generally be promised.

Since the optimization problem of electrical machines belongs to the multi-objective nonlinear mathematical programming problem, and each objective is normally a nonconvex function with multiple extreme points, the stochastic algorithms or also named intelligent algorithms with the ability of global convergence are widely utilized. First of all, various scalar or single objective intelligent optimization algorithms have been employed for the optimization of electrical machines in the past several decades, such as the evolutionary algorithms (EAs), particle swarm optimization (PSO) algorithms, estimation of distribution algorithms (EDAs), immune algorithm, and ant colony algorithm, and their improvements [22-26]. For the widely used genetic algorithm (GA) and differential evolution algorithms (DEAs), they are two of four major subclasses of EAs. The other two subclasses are evolution programming and evolution strategy. Even though with the ability of global convergence, the convergence speed is slow compared with the gradient-based algorithms since various iteration steps are required for global optimum searching.

To solve the multi-objective problem, the objectives are normally summed in one objective function with the weighting factor or reflected in the constraints. Furthermore, multi-objective optimization algorithms can be applied directly. They can provide a Pareto solution set with a single run. This solution set consists of many non-dominated optimal design solutions for the designer to select based on a specific application (meaning specific weighting factors). Some popular multi-objective optimization algorithms are originated from GA and PSO[27, 28], such as multi-objective GA, non-dominated sorting GA (NSGA) and its improvements (NSGA II) [29, 30], and multi-objective PSO algorithm [31- 33]

2.6.2 Comparison and Comments

The gradient-based algorithms show faster convergence speed but poorer global convergence ability compared with intelligent algorithms. The sensitivity of the objectives and constraints with respect to the design variables are required for the application of the algorithms. Based on the general motor modeling method such as FEM without explicit analytical expression, the sensitivity analysis is complex. If a differential method is applied, the number of the FEM for each iteration should be conducted is twice of the number of the design variables which increases the computation burden from the aspect of performance calculation. When some fast analytical or coarse models are applied for the optimization, the computation time for a single sample can generally be ignored. Considering the above-mentioned reasons, the application of pure gradient-based algorithms is relatively less. The utilization of gradient-based optimization can be found in the design problem of a very high dimension such as topology optimization.

Different from conventional gradient-based algorithms, the implementation of intelligent optimization algorithms does not depend on the mathematical properties of the optimization models. For example, the functions of objectives are not required to be continuous and differentiable, and to have analytical expressions. Compared with conventional ones, intelligent optimization algorithms have many advantages, such as global optimizing, stronger robustness and parallel searching capability. There are three main steps in the implementation of intelligent optimization algorithms. They are population initialization and algorithm parameter determination, individual updating and population regeneration. Different algorithms have different algorithm parameters and strategies for individual updating and population regeneration. However, the convergence speed is normally low for these algorithms because of the huge amount of iterations.

Moreover, we can see the works for improving both types of algorithms by taking advantage of each other i.e. the hybrid algorithms. For example, the gradient algorithm with stochastic multi-start points and the intelligent algorithm with gradient information for the solution updating.

For the conduction of the optimization for electrical machines, besides selecting rational performance calculation model and optimization algorithms, the optimization methods or strategies are also important for improving the optimization efficiency. Therefore, several optimization methods/strategies will be discussed in the next session.

2.7 Techniques and Strategies for Optimization Acceleration

The high computation burden is always the problem for the conduction of the optimization of electrical machines. As introduced above, the computation cost mainly comes from performance calculation, a huge number of samples required for the high-dimensional problem, and the iteration required for optimization. Taking a ten-parameter optimization problem as an example, if an intelligent algorithm like GA is applied as the optimization algorithm, the total sample size can be roughly estimated by 10*5*200, where 10*5 is the population size of each iteration step, and 200 is an average number of iteration. The performance calculation of each sample can be time-consuming if the modeling method of high accuracy is applied such as FEM.

In addition to the computation power improvement such as the hardware update, model reduction, and optimization algorithm enhancement, the strategies or methods on the reduction of computational efforts are accordingly proposed for relieving the pressure in these aspects, which include application of surrogate model, design of experiment (DoE), reduction of design space, sequential optimization method, and space mapping method, etc.

2.7.1 Surrogate Models, Modeling Techniques and Optimization

2.7.1.1 Surrogate Models

There are several types of surrogate models, such as response surface model (RSM)[34, 35], radial basis function (RBF) [36-38] model, Kriging model [39-41], artificial neural network model (ANN) [42,43]. Meanwhile, some improvements have been developed form them in the last several decades, such as the moving least square model (MLSM, an improvement of RSM) and compactly supported radial basis function model (an improvement of RBF model). All of them have been employed for the optimization design of electromagnetic devices and systems including electrical machines. Three widely used ones, RSM, RBF, and Kriging models, are compared in Table 2-1. In the equations, y is the response for an input $\mathbf{x}$, $\{\mathbf{x}_1, \mathbf{x}_2, \ldots, \mathbf{x}_n\}$ are n given samples, Y is a matrix of the responses of the samples, X, H, and Q are matrixes of the basis functions, R is a matrix of the correlation functions.

Table 2-1 Comparison of RSM, RBF, and Kriging models

Model	RSM	RBF	Kriging
Equation	$y = X\beta + \epsilon$	$y = \displaystyle\sum_{i=1}^{n} \beta_i H(\|\mathbf{x} - \mathbf{x}_i\|)$	$y = q(\mathbf{x})'\beta + z(\mathbf{x})$
Explanations	β: coefficient matrix X: structural matrix ϵ: a stochastic error	β_i: coefficient matrix $H(\cdot)$: RBF function $\|\mathbf{x} - \mathbf{x}_i\|$: Euclidean norm	β: coefficient matrix $q(\mathbf{x})$: basis function $z(\mathbf{x})$: a stochastic process with mean of zero, variance of σ^2, and covariance related to the correlation function matrix.
Popular model basis functions	Linear and quadratic polynomials	Gauss, multiquadric, and inverse multiquadric	Constant, linear and quadratic polynomials
Correlation functions	n/a	n/a	Gauss and exponent
Parameter estimation	$\hat{\beta} = (X'X)^{-1}X'Y$	$\hat{\beta} = H^{-1}Y$	$\hat{\beta} = (Q'R^{-1}Q)^{-1}Q'R^{-1}Y$ $\hat{\sigma}^2 = \dfrac{1}{n}(Y - Q\hat{\beta})'R^{-1}(Y - Q\hat{\beta})$
Estimation method	Least square method	n/a (no error term)	Best linear unbiased (for β), and maximum likelihood (for σ^2) estimation methods
Modeling features	Global trend (mean response)	Global trend	Global trend and local deviation
Model complexity	Low	Middle	High

Mathematically, RSM and RBF are parametric models, while the Kriging model is a semi-parametric model as it consists of two terms, deterministic term and stochastic process term. Compared with the RSM and RBF model, Kriging is superior in the modeling of local nonlinearities [43].

Considering the modeling complexity, RBF is higher than RSM because there are extra shape parameters in the model basis functions in RBF while RSM does not have these parameters. The Kriging model is the most complex one. As shown, not only the coefficient matrix in the deterministic term but also the parameters in the stochastic process term like variance and parameters in the correlation function have to be estimated in this kind of model. For this purpose, the best linear unbiased estimation (BLUE) method is employed for the estimation of β, and the maximum likelihood estimation (MLE) method is used to estimate variance σ^2. To effectively estimate these parameters, some software packages such as Design and Analysis of Computer Experiments (DACE) can be found coded by MATLAB [41]. More details of these models and their applications can be found in references [38-40, 42].

2.7.1.2 Modeling Techniques

To obtain an accurate surrogate model, modeling techniques are required to investigate. Normally, there are several steps for the development of a surrogate model, including generation of samples by using DoE techniques, model construction and verification [35, 44-46]. Figure 2-2 illustrates the main steps, which are briefed as follows.

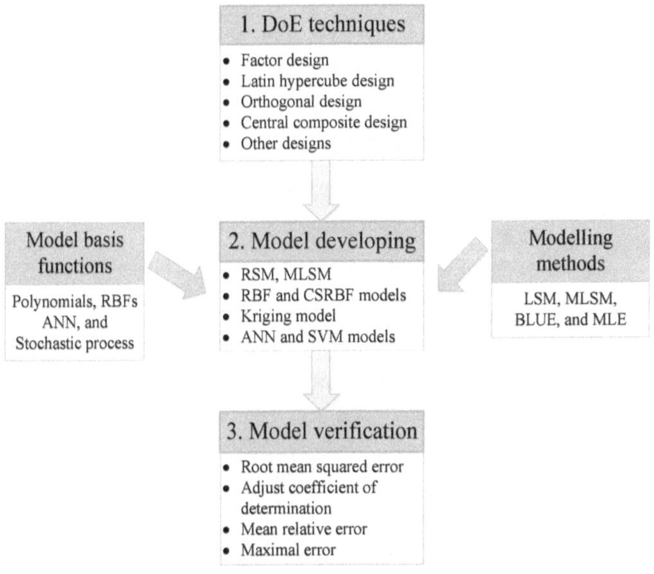

Figure 2-2 Design flowchart for surrogate models.

Firstly, generate some samples by using a DoE technique. There are many available DoE techniques in statistics, such as full factor design, central composite design, and orthogonal design.

Secondly, develop a kind of surrogate model based on the obtained samples. Table 2 lists the parameter estimation methods and solutions for RSM, RBF, and Kriging models. More details can be found in reference [4]. Many DoE techniques and surrogate models have been integrated into the software, such as MATLAB, SPSS, and MINITAB.

Finally, to ensure that the constructed model is accurate, error analysis should be conducted based on some new and different samples. Several criteria may be applied for this purpose, such as root mean square error and mean relative error. If the model is with good accuracy, it can be used for the following optimization, and the obtained optimization results are reliable.

2.7.1.3 Comments

First, there are two ways for the implementation of surrogate models in the optimization. Assume the optimization objective is maximizing the motor efficiency $\max: f(\mathbf{x}) = \eta$ and FEM is employed as the electromagnetic analysis model to obtain parameters back electromotive force (EMF), winding inductance and core loss. The first way is to develop a surrogate model like RSM for the efficiency directly. The second way is to develop three surrogate models to replace the FEMs used for the evaluation of three electromagnetic parameters, back EMF, winding inductance and core loss. Then motor efficiency is calculated based on those three models. Though the required FEM samples are the same for the two ways, the latter always has higher accuracy than the former. The main reason is that all of them are nonlinear to the design parameters. The accumulated error is higher if we incorporate them into one function.

Second, as compared in Table 2-1, the Kriging model has better modeling capability of local nonlinearities while it is more complex compared with RSM and RBF. However, Kriging has attracted more attention recently as there are no significant differences in the computation speeds of these models in personal computers and a public toolbox is available.

Third, there are two main issues for the application of surrogate models. They are the computation cost of DoE samples and model accuracy. Full factor DoE techniques have been used to increase the accuracy of the solution in many situations. However, they require high FEM simulation costs, especially for high-dimensional situations. For example, considering a PM motor with 5 parameters for optimization, 10^5=100,000 FEM samples are needed if a 10-level full factor DoE technique is applied, and 5^5=3,125 FEM samples are needed if a 5-level full factor DoE technique is applied for the development of RSM or Kriging. The model with a 10-level full factor DoE technique has higher accuracy than the 5-level one; however, the computation cost is huge for most designers.

Therefore, optimization efficiency is a big issue for both methods. To attempt this issue, some new and efficient optimization methods are introduced as follows. The first one is a Sequential Optimization Method (SOM) based on a space reduction method, which is mainly proposed for low-dimensional design problems. The second one is a multilevel optimization method, which is proposed for the high-dimensional design problems, for example, the dimension is larger than 10. The last one is the Space Mapping (SM) method.

2.7.2 Sequential Optimization Method Based on Space Reduction

2.7.2.1 Method and Flowchart

The large design space of electrical machines is due to the high dimension of the design variables. However, the optimal solution only locates in a small subspace (can be called the interesting subspace). If this fact is ignored, a lot of the samples outside the interested subspace will be evaluated during the optimization, resulting in an unnecessary waste of computation cost. The idea of the SOM is to reduce this kind of waste by using a space reduction method.

Figure 2-3 shows a brief framework for the space reduction method. Based on this method, SOM consists of two main processes, design space reduction and parameter optimizing. The main aim of the former process is to efficiently reduce the initial design space to the interesting subspace as shown in Figure 2-3 (b) by using several space reduction strategies. The target of the latter process is to pursue the optimal solution in the small subspace [47, 48].

This kind of optimization strategy can be applied to the multi-objective optimization situation. However, the main difference and challenge, in this case, are the shape of interesting space. It is a manifold rather than a normal square or a cube. Consequently, a new space reduction method is required. A multi-objective SOM based on a modified central composite design (a kind of DoE technique) was presented for the design optimization of electrical machines [49, 50]. Through the study on several examples including design optimization of a PM-SMC transverse flux motor (TFM), it was found that the required FEM samples of single- and multi-objective SOMs are around 10% of those required by direct optimization method (for example, DEA plus FEM). Thus, both methods are efficient for the optimization of electrical machines [49, 50].

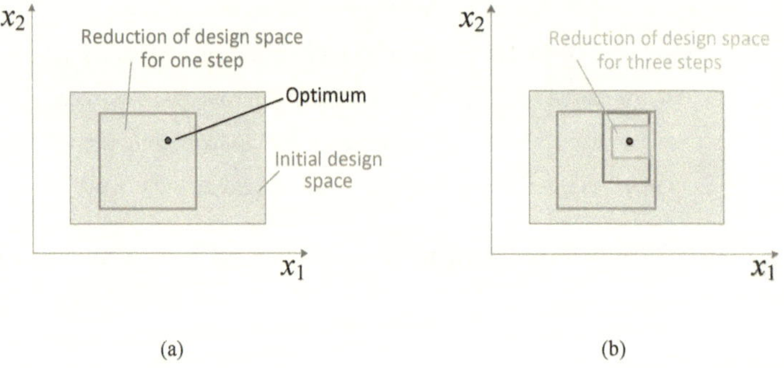

Figure 2-3 Illustration of space reduction method in the SOM, (a) one step, (b) three steps.

2.7.2.2 Comments

First, similar to the direct and in-direct (surrogate model based) optimization methods, the proposed SOM has difficulty for high-dimensional design optimization problems due to the large amount computation cost of FEM. The main reason is that this kind of method does not reduce the design space or parameters. Therefore, this method only benefits the low-dimensional design optimization problems (dimension less than 5).

Second, for surrogate model based optimization, this method may be efficient for higher dimension situations (for example dimension is 5 or 6) if partial DoE techniques like orthogonal design are applied. It is found that DoE is key for the efficiency of the SOM and partial DoE techniques can be used to improve the efficiency of SOM [48].

Third, the efficiency of SOM does not highly dependent on the type of surrogate models. Though Kriging is claimed superior to RMS and RBF in many studies, there are no significant differences in the objective when they are applied in the SOM. The main reason is the SOM is a kind of iterative optimization, and the differences can be decreased in the sequential optimization process.

2.7.3 Multilevel Optimization Method

2.7.3.1 Method and Flowchart

It is noted that the design parameters have different sensitivities in terms of design objectives, meaning that some parameters are sensitive to the objectives and others are not sensitive. Therefore, more attention should be paid to the sensitive parameters in the optimization.

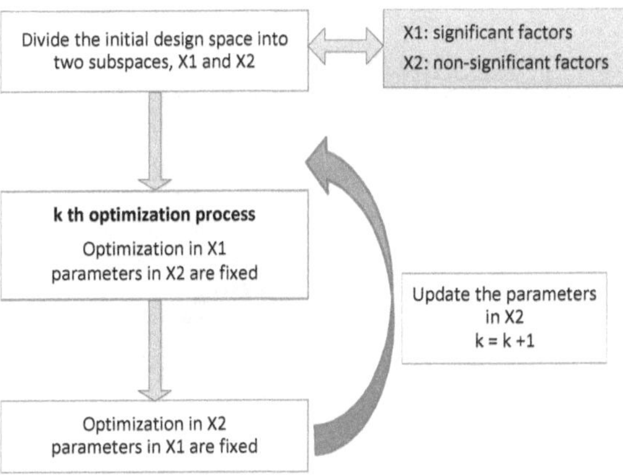

Figure 2-4 Main flowchart for the multilevel optimization method.

Based on this idea, a multilevel optimization method as shown in Figure 2-4 has been developed and successfully applied to the successful design of electrical machines [51-54]. As shown, the first step in the multilevel optimization is to divide the initial large and high-dimensional design space into several low-dimensional subspaces by using sensitivity analysis (SA) methods [51]. A structure with two subspaces is shown in Figure 2-4. In the implementation of optimization, the subspaces will be optimized sequentially regarding the sensitivity orders of the covered parameters. Thus, the parameters with higher sensitivities will be optimized before those with lower sensitivities. The main advantage of this method is that each subspace is low-dimensional and all the general optimization methods can be employed with good efficiency.

2.7.3.2 A Case Study

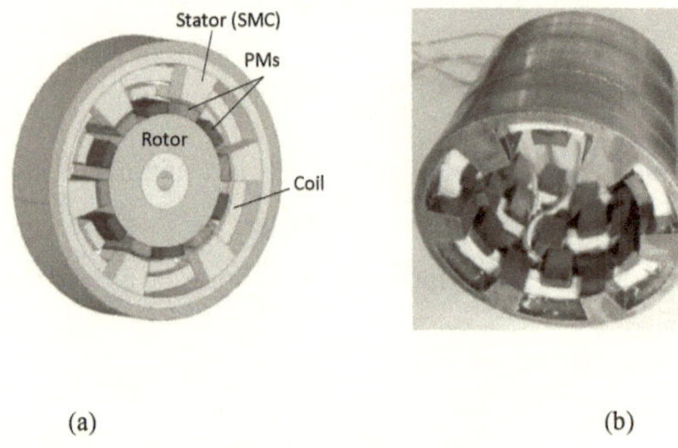

Figure 2-5 Structure of the CPM (a) and its SMC stator prototype (b).

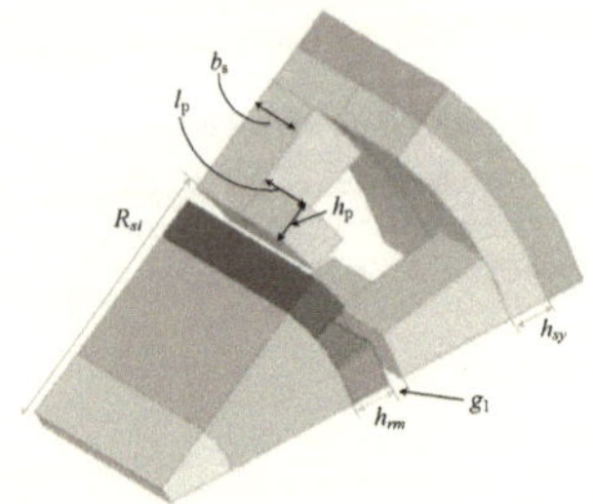

Figure 2-6 Main dimensions of the CPM.

This case study investigates an optimization problem of a small PM claw pole motor (CPM) with cores of soft magnetic composite (SMC) material for domestic applications like washing machines. Figure 2-5 shows the structure of the motor and a prototype of the SMC stator. Based on our design experience, nine parameters including seven dimension parameters shown in Figure 2-6 and the number of turns of the winding (N_c) and SMC core density (ρ) are important for the performance of this motor [51]. Thus, they will be considered as the design optimization parameters for this case study. Table 2-2 lists their values for the prototype of the SMC stator.

Table 2-2 Main dimension and design parameters of the CPM prototype

Par.	Description	Unit	Value
---	Number of phases	---	3
---	Number of poles	---	12
R_{so}	Stator outer radius	mm	33.5
R_{si}	Stator inner radius	mm	21.5
R_{shai}	Radius of shaft hole	mm	2.5
b_s	Width of side wall	mm	6.3
h_{rm}	Radial length of magnet	mm	3
ρ	SMC core's density	g/cm^3	5.8
g_1	Air gap	mm	1
h_p	Claw pole height	mm	3
h_{sy}	Stator yoke thickness	mm	3
l_p	Axial length of claw pole	mm	5.8
N_c	Number of turns of winding	turn	256

The optimization aims to minimize the active material cost while maximizing the output power. Constraints include efficiency (η), output power (P_{out}) and current density of the winding (J_c). The optimization model can be defined as follows.

$$\min: \quad f(\mathbf{x}) = \frac{\cos t}{c_0} + \frac{P_0}{P_{out}}$$
$$\text{s.t.} \quad g_1(\mathbf{x}) = 0.78 - \eta \le 0 \qquad (2\text{-}2)$$
$$g_2(\mathbf{x}) = 60 - P_{out} \le 0$$
$$g_3(\mathbf{x}) = J_c - 4.5 \le 0$$

where C_0 and P_0 are the cost and output power of the initial prototype.

As introduced above, for the implementation of the multilevel method for the nine-parameter problem, there are two steps, i.e. determination of multilevel structure and optimization. To divide the whole design space of the high dimension into subspaces of low dimension, sensitivity analysis methods are normally required. Three popular approches are the local and global SA and analysis of variance (ANOVA) based on the DoE technique [51]. Table 2-3 tabulates the results for the sensitivity analysis of these three methods. For ANOVA, it uses * instead of value to show the significant factors. Moreover, as the number of turns of winding does not include in the FEM, it is not shown in this table.

As shown in Table 2-3, there are significant numerical differences in the parameter sensitivities provided by the local and global SA methods. However, the highest four parameters are the same, and they are ρ, b_s, h_{rm} and R_{si}. For a good balance of the subspaces, these four parameters can be grouped into X1. Then the other four parameters will be placed into X2. For the ANOVA results based on orthogonal DoE, it indicates that there are three significant parameters (b_s, ρ and h_{rm}), which are also the highest three parameters of local and global SAs. Therefore, there are no significant differences between these methods.

Table 2-3 Sensitivity analysis results for the CPM

Par.	Local SA	Global SA	ANOVA
R_{si}	0.0267	0.0267	
b_s	0.1095	0.0990	*
h_{rm}	0.0754	0.0876	*
ρ	0.1441	0.1587	*
g_1	0.0141	0.0171	
h_p	0.0053	0.0102	
h_{sy}	0.0019	0.0026	
l_p	0.0222	0.0131	

The optimization is then conducted for each subspace sequentially with effective methods. Table 2-4 lists the optimization results by employing Kriging as the surrogate model and DEA as the optimization algorithm for each subspace optimization. As shown, the motor performances have been improved significantly by multilevel optimization compared with the initial design. In detail, motor output power has been increased by 171% (163 vs 60 W), the active material cost has been decreased by 35% ($14.18 vs 9.17). Considering the computation cost, the FEM samples have been decreased to 6.71% compared with the samples required by the direct optimization method. The huge computation cost of the direct optimization method with FEM makes it computationally prohibitive for processing on a personal computer. Based on our experience, it is hard to claim that its solution is superior to the one obtained from the multilevel optimization method as this is a high-dimensional and strongly nonlinear optimization problem (meaning that there are many local minimal points). Therefore, the multilevel optimization method is efficient for high-dimensional problems.

Table 2-4 Optimization results of the CPM

Par.	Unit	Initial design	Multilevel optimization	Direct optimization (FEM + DEA)
η	%	78.0	82.4	---
P_{out}	W	60	163	---
$Cost$	$	14.18	9.17	---
FEM samples	---	---	604	9,000

2.7.3.3 Comments

The multilevel strategy transfers the computationally prohibitive high-dimension optimization into a computable low-dimension problem. For the optimization in each subspace, techniques such as the surrogate model can be applied for accelerating the optimization further.   For the application of the multilevel method, the effective SA and rational division of the subspaces are the key factors for the optimization implementation.

2.7.4 Space Mapping Method

2.7.4.1 Method and Flowchart

The space mapping method is another optimization method proposed for the design optimization of electromagnetic devices, and it can be applied to the design of electrical machines. Figure 2-7 shows a brief mapping framework for this method. As shown, two spaces are investigated for a specific design problem. They are fine and coarse model spaces. For electrical machines, the fine analysis model can be a finite element model or an analytical model; the coarse model can be a magnetic circuit model or a surrogate model. The optimization is conducted in the coarse model space only. However, the optimization solution is not good due to the less accuracy of the coarse model. Therefore, several optimization loops may be required in the implementation of the SM method. To guide the searching direction of SM optimizing, a fine model and a mapping algorithm are used. Besides the original SM, several improvements have been developed, such as the aggressive and implicit SM methods. These methods can be applied to the design optimization of electrical machines [55-56].

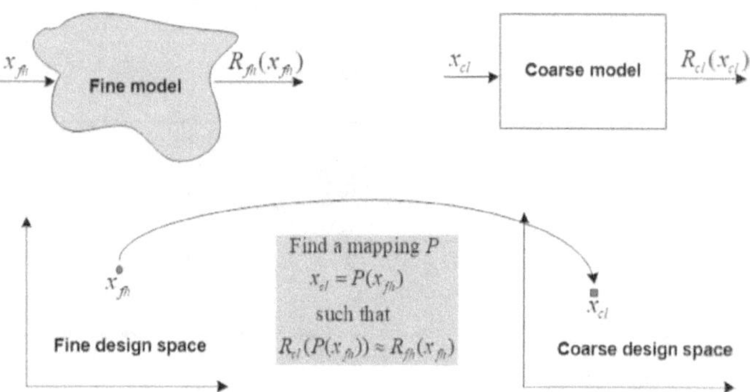

Figure 2-7 Framework of space mapping optimization method.

2.7.4.2. Comments

The main differences between the multilevel optimization methods and the SM method are discussed as follows.

First, there are two requirements for the optimization in the coarse design space of SM methods, fast optimization efficiency and acceptable accuracy. Therefore, the magnetic equivalent circuit model is a good candidate for the simulation of coarse design space. Surrogate models may be suitable for the low dimensional situation. For high dimensional situations, surrogate model based optimization should be integrated with multilevel optimization strategy to improve the optimization efficiency.

Second, the mapping algorithm is critical for the SM methods. It will link the two spaces and use the responses in the fine design space to guide the optimization direction in the coarse design space. However, there is only one design optimization space in the multilevel optimization method. And the key factor of the multilevel optimization method is the determination of the subspaces.

2.8 Rethinking, Challenges, and Proposals

As summarized above in 2.2 and shown in Figure 2-8, the main process of the design optimization includes the initial structures proposing, quantitive design, modeling and optimization, and manufacturing for validation and production.

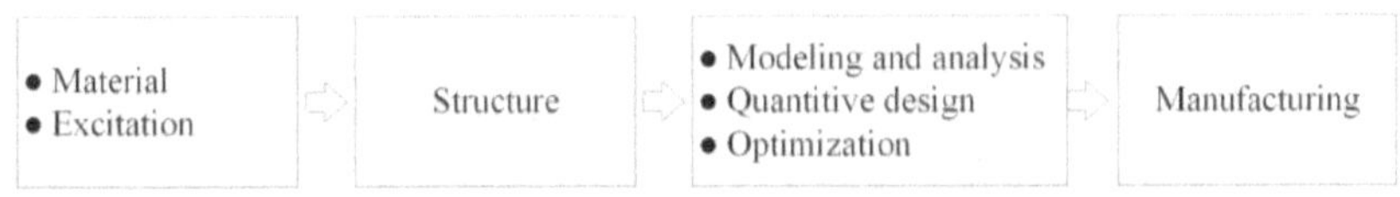

Figure 2-8 Diagram of the basic modules in the design and optimization of electrical machines.

On one hand, we should have seen the development of any single module such as the development of novel materials, structures, drive, and control methods for excitation, accurate modeling method, above reviewed efficient optimization techniques, and novel manufacturing technique, etc. contribute to the advance of the electrical machines. On the other hand, a more integrated design optimization process considering each module with high freedom, automation and design optimization for electrical machines is what we always pursued.

Starting with the optimization module, we can find that the techniques used in the optimization module are based on the condition of given initial structure designs. Generally, the given of the initial design is highly dependent on the designer's expertise on the motor structure, material, and excitation methods, etc. The design optimization process contains the selection of the known structures or even the bring-up of new structures and then the parametric optimization is conducted for the optimal solution according to the objectives and constraints. The limitation of this process is the dependence on the designers' experience in the initial design development and the low optimization freedom of the parametric optimization for a given design. Once the initial design is given, the following optimization may be already conducted in a suboptimal domain. The low freedom optimization with several design variables such as the size and material property restricts the optimization further in obtaining the optimal design. Therefore, how to effectively complement the drawback on the initial structure design and overcome the limitation of the parametric optimization for high freedom design optimization should be investigated.

If we considered the pursuing of the integrated design optimization is the upward compatible process of the optimization module, the consideration of the manufacturing influence on the performance of the electrical machines in the optimization can be regarded as the down compatible process. If the impact of the manufacturing such as the tolerances on the reliability of the production, and manufacturing costs, etc. are considered as a branch of the multidisciplinary analysis, the whole process from design optimization to production can be more integrated, complete and the development cycle can be accelerated.

Furthermore, we can see that the electrical machine exists as one of the components of the drive or electricity generation system. A specific example is the drive system design of electric vehicles which include batteries, electrical machines, drive circuits, and transmission devices, etc. The optimal design of the system is never limited in the single-component optima but the system level. The application-oriented system-level problem may change the design optimization space and produce more comprehensive objectives, constraints for the optimal system. The application-oriented system-level design optimization of electrical machines and their drive systems can be regarded as the unification of the techniques for the system design optimization problem.

For the above-noted challenges, the proposals and practices are discussed below which include topology optimization for electrical machines, design optimization for production, and application-oriented design optimization for electrical machines.

2.9 Topology Optimization for Electrical Machines

Structure "is an arrangement and organization of interrelated elements in a material object or system, or the object or system so organized" while the excitation in the electromagnetic field is the process of establishing the magnetic field by the current. The structure of the electrical machine is an effective combination of the materials and excitation in space and time. **Therefore, we consider the design optimization of**

electrical machines as the process of searching for the optimal spatial and temporal distribution of the materials and current excitation.

To find the optimal layout of material in the design domain, topology optimization methods are developed in the mechanical field firstly for the light-weight design of mechanical structures in the seminal paper by Bendsøe and Kikuchi in 1988 [57].

As presented in Figure 2-9 (a simplified demonstration), this topology optimization presents a different process compared with the parametric design optimization method such as size optimization and shape optimization. Sizing optimization has been recognized as the most direct optimization approach which aims to reach the optimal objectives by changing the size parameters. However, since the shape or the topology of the initial design is already fixed, the improvement of the performance is usually limited.

Compared with the size optimization method, the shape optimization approach may be more effective, which usually takes the geometrical parameters of some boundary points as design variables. Then performance can be optimized by changing the shape of the design. Nevertheless, the limitation in the shape development still exists. For example, holes cannot arise unrestrained. Moreover, the selection of the boundary points is usually based on experience or analytical calculation which increases the difficulty of the realization.

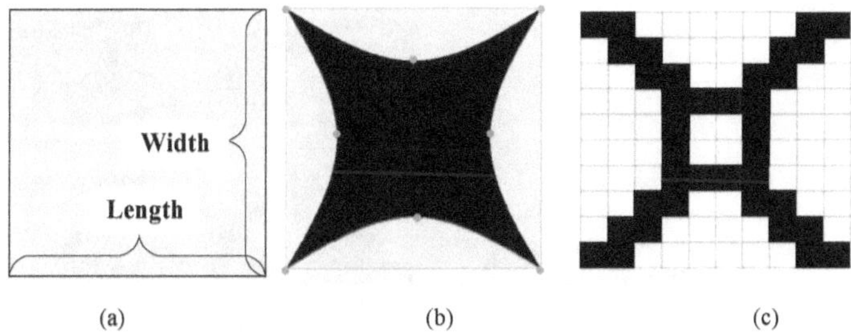

Figure 2-9 Schematic diagram of different optimization methods: (a) size, (b) shape, and (c) topology.

Compared with the two approaches, topology optimization attempts to find the optimal layout of the structure for the objective performance. Currently, this method takes advantage of the CAE analysis method such as FEM and transferring the structure optimization problem in the continuous design space into a finite high dimensional mathematical programming problem. Taking the topology optimization based on FEM as an example, the material such as ferrite magnetic material and air of each element is arranged by the optimization process. From this point of view, topology optimization integrates the conceptual design proposal and optimization into an automatic process.

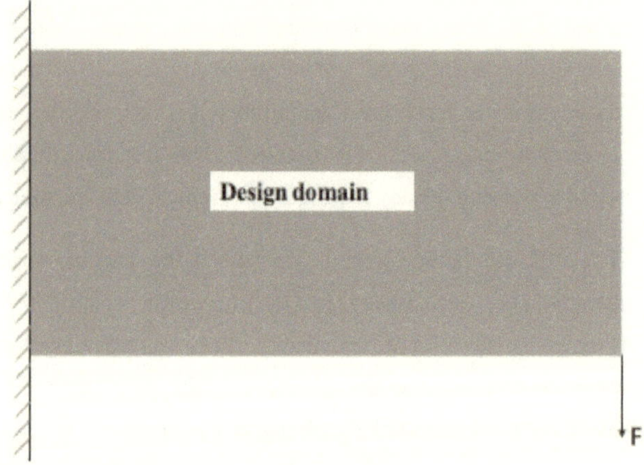

Figure 2-10 Cantilever beam design problem.

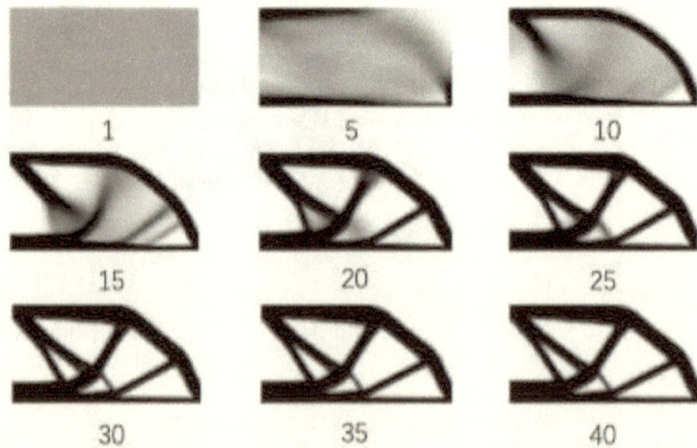

Figure 2-11 Shape variation in different iteration steps.

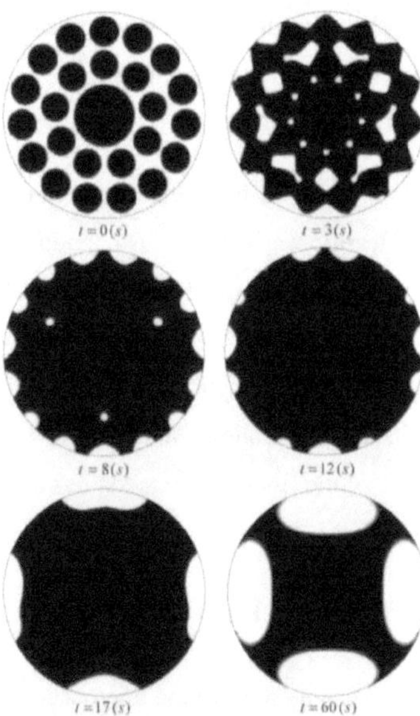

Figure 2-12 Rotor shape variation v.s. iteration time.

The thought can be illustrated by the simple mechanical example of the design of a cantilever beam by topology optimization. As shown in Figure 2-10, the problem can be described as the optimization for maximum stiffness with the required volume of solid material under the given loading **F** and the boundary condition [59]. FEM is adopted as the modeling method for the example and the material property (solid or air) of each element is set as the design variable. The following Figure 2-11 presents the iteration steps of the topology optimization by density method which will be introduced in the following Chapter 5. After the optimization iteration, the novel conceptual structure design is obtained where the black part means the solid material and the white part means the air.

Due to the advances of the ideology, topology optimization techniques have been initiated and researched extensively in the mechanical field and also extended to other engineering fields such as the thermotic, acoustics, and electromagnetics, etc. [60]. A design optimization example of the rotor of a reluctance motor by topology optimization method can be found in reference [61]. Figure 2-12 presents the rotor shape variation in the optimization iteration. By the topology optimization method, the rotor of the reluctance motor is designed and optimized automatically for by the ferromagnetic material in the design space with high freedom.

However, due to the complexity of the design of electrical machines such as the various materials, nonlinearity, variation of excitation, etc., the development of topology optimization methods for electrical machines is still at the early stage. The dependence on the initial design is yet eliminated. Nevertheless, the advantages of topology optimization are obvious. The optimization freedom on shape-changing is highest among the size and shape optimization. Since the research on topology optimization is an important chapter of the Book, the detailed review of the current topology optimization methods for electrical machines is presented in Chapter 5.

2.10 Robust Optimization for Electrical Machines

The deterministic design optimization approaches do not take account of the inevitable uncertainties in practical manufacturing, which contains but are not limited to material variations, manufacturing tolerances and operation conditions. The research on the influence of uncertainties on the performance of electrical machines such as the cogging torque and back EMF can found in [62-65].

The deterministic optimal solution generally locates at the boundary conditions of the optimization model and the performance may sensitive to the uncertainties. Therefore, the performance of the actual operation of the optimal design will degenerate or even fail because of violating the constraints. From the aspect of batch production, there is also the risk of high-performance diversity. From this perspective, the robustness of the solution should also be considered as an important index for the final solution. Consequently, the robust optimization research aims to model the influence of the uncertainties in the optimization process and make the trade-off between the objective performance and the robustness until obtaining the optimal robust solutions.

The Taguchi method is regarded as one of the early efforts on the robust optimization which is named after Genichi Taguchi, who is regarded as the "father of quality engineering". The signal to noise ratio is defined to quantify the influence of the noise factors on the performance.

$$SNR := -10\log_{10}\left(\frac{1}{N}\sum_{i=1}^{N} f\left(\mathbf{x},\xi_i\right)^2\right) \tag{2-3}$$

where $\mathbf{x}$ is the design parameter and ξ_i is the noise factor. Even though the method is controversial mathematically, it achieved widely application due to its practicability [66-68] for robust design.

With the development of uncertainty quantification and robust optimization techniques [69], there is a widely used classification of the uncertainty which divides it into aleatory and epistemic uncertainties. Aleatory uncertainty is generally the inherently physical nature and described with stochastic variables. Epistemic uncertainty is caused by the shortage of recognition and information, which can be expressed with interval variables or fuzzy sets.

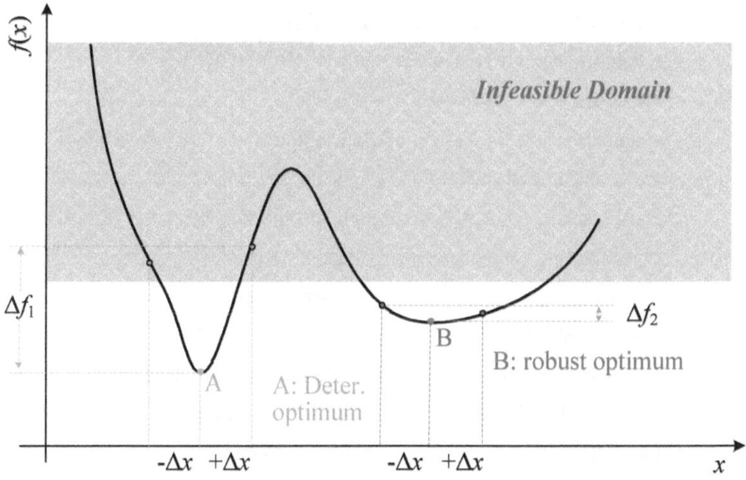

Figure 2-13 Schematic diagram of the RDO and RBDO with uncertainty.

Another discussion about the classification of the uncertainties can be found in [70] which divides the uncertainties into four categories, i.e.: **Changing environmental and**

operating conditions: This kind of uncertainty is an inherent property of most systems. As introduced before, the design of electrical machines is a multidisciplinary problem. Different operation statuses including but not limited to environment temperature, drive methods and loading conditions, etc. should be considered in the design. **Production tolerances and actuator imprecision:** Due to the manufacturing tolerances, the exact design parameters cannot be realized in the practical engineering which leads to the fluctuation of the objective performance. **Uncertainties in the system output:** This type of uncertainties come from the errors of modeling method and measurements. **Feasibility uncertainties:** This kind of uncertainty is about the uncertainty impact on the constraints. The methods for quantification of the uncertainties mentioned above are summarized as the deterministic method, stochastic method, and possibilistic method. Four kinds of uncertainties and three types of quantification methods make up 12 concepts of the robustness.

The research on the second and fourth types of uncertainties are generally focused, i.e. uncertainties that influence the objectives and constraints. Accordingly, there is the classification of optimization method i.e. robust design optimization (RDO) and reliability-based design optimization (RBDO). RDO aims to find the optimal solution of the deterministic performance which is also insensitive to the uncertainties. As a unified example to show the RDO and RBDO, Figure 2-13 illustrates a one-dimensional problem, where the dashed lines are used to demonstrate the fluctuation intervals of points A and B. As shown, point A is the minimum objective. However, the sharp variation of the objective value in the uncertainty interval means high-performance diversity of the same nominal design, which should be avoided in the design. On the other hand, the suboptimal point B shows a lower performance fluctuation or higher robustness to uncertainties which may be selected as the robust optimal solution. RBDO considers the feasibility of the optimal design which proposes to decrease the risk of the optimal solution violating the performance constraints. As shown in Figure 2-13, the robust solution should be in the feasible constrained feasible region under the uncertainties.

In summary, the uncertainty quantification for both optimization methods is basically modeling its effects on the performances in the optimization model. We further summarized them into one kind of uncertainty. Correspondingly, the RDO and RBDO can be classified into one generalized category.

Therefore, in this survey, the classification and discussion on the uncertainty and robust optimization are based on the uncertainty quantification methods which is divided into two categories i.e. stochastic and non-stochastic methods for the relevant type of uncertainties. In addition to the different uncertainty quantification methods, the other factors that influence the effectiveness of robust optimization such as the strategies are also concerned.

2.10.1 Robust Design Optimization Methods with Stochastic Uncertainties

For objective performances, the robust measures with stochastic uncertainties take advantage of the stochastic property of the uncertain variables to minimize the mean and variance or standard deviation of the objective functions. Considering the performances in the constraints i.e. feasibility of the solution, the large percentage of the robust design should be promised to meet the requirement of the constraints. The stochastic feasibility is generally defined as

$$P\left(g_i(x) \leq 0\right) \geq P_i', i = 1,\ldots,m \tag{2-4}$$

where P_i' is the desired probability. Another method is utilizing the moment information of the constraints i.e. the first and second moments (or mean values and variance values) which is known as moment matching method. The constraints in the optimization model are built as

$$\mu_{g_i} - k_i \sigma_{g_i} \geq 0 \tag{2-5}$$

where μ_{g_i} and σ_{g_i} are the mean of the standard deviation of the constraints respectively, and k_i is the sigma level. The high feasibility of the design can be obtained by adjusting the sigma level. The design for six sigma (DFSS) techniques is applying this modeling.

Monte Carlo approach (MCA) has been widely used in the mean and variance information calculation of the objectives and constraints of the objective and constraints, and reliability calculation in (2-4). However, the accuracy of MCA depends on the sampling number, the error of the MCA is inversely proportional to the square root of the sample number. Generally, a huge amount of samples is required. Due to the model complexity for the performance evaluation by a general method such as FEM, the direct application with MCA with FEM cannot be applied because of the heavy computation burden. Some works of the author's research group take advantage of the Kriging model for modeling of the performances in the whole design domain. Then the computation burden of MCA is effectively decreased [71, 72]. The following two case studies are basically based on this method. Since the robust solution can only be the local or global optima or boundaries, the robustness analysis is only necessary for these potential solutions instead of middle solutions. In addition to the computation burden reduction by the aforementioned robustness measurement method, it has been mentioned that computing cost can be relieved further by reducing the solution required to do the reliability analysis [73]. Some other works based on the Taylor series can be found in [74, 75] for calculation of feasibility. However, the establishing of the Taylor extension requires the sensitivity information of functions with respect to the uncertain variables which increases the difficulty of application. The stochastic response surface methods offer an effective way for the effective evaluation of moment information [73, 76-79]. Taking PCE as an example, only very limited samples are required for the PCE establishment. Once the model is built, the relevant mean and variance of the performance can be calculated directly by the model coefficients.

From the point of the feasibility of constraints, the most probable point (MPP) based method received much attention in the reliability analysis for RBDO. By applying the MPP based method, the most probable failure point is searched firstly, then to calculate the reliability of constraints and find the optimal solution subject to the reliability requirement [80-85].

2.10.2 A Case Study Based on DFSS for Stochastic Uncertainties

DFSS is the engineering application of the moment matching method for stochastic uncertainties. The uncertainties in the manufacturing vary in a small range. In this case, the objective and constraint functions are generally linear with respect to the uncertain variables. The normal distribution is one of the most common uncertainty distribution forms in practical engineering. Assume the design variables are also with the uncertainty follows the normal distribution with different means (μ) and standard deviations (σ). Therefore, all constraints and objectives are variables now and can be formulated as functions of mean and standard deviations. Supposing the design variables are also with uncertainties (suitable for the case below), the robust optimization model under DFSS can be expressed as

$$
\begin{aligned}
\min: \quad & F\left[\mu_f(\mathbf{x}), \sigma_f(\mathbf{x})\right] \\
\text{s.t.} \quad & g_j\left[\mu_f(\mathbf{x}), \sigma_f(\mathbf{x})\right] \le 0, j = 1, 2, \ldots, m \\
& \mathbf{x}_l + n\sigma_x \le \mu_x \le \mathbf{x}_u - n\sigma_x \\
& \text{LSL} \le \mu_f \pm n\sigma_f \le \text{USL}
\end{aligned}
\tag{2-6}
$$

where LSL and USL are the lower and upper specification limits, n is the sigma level, which can be equivalent to short-term and long-term defects per million opportunities (DPMO) as shown in Table 2-5. As shown, the 6-sigma level is equivalent to 3.4 DPMO in the long-term quality control, which is a high standard for industrial production and has been widely adopted by many enterprises worldwide for the quality control of the products. MCA is widely utilized to evaluate the μ and σ of objectives and constraints, and the sigma levels of constraints [86].

Table 2-5 Defects per million opportunities regarding sigma level

Sigma level	DPMO (short term)	DPMO (long term)
1	317,400	697,700
2	45,400	308,733
3	2,700	66,803
4	63	6,200
5	0.57	233
6	0.002	3.4

Table 2-6 Main dimensions and parameters for the TFM

Par.	Description	Unit	Value
---	Number of stator teeth	---	60
---	Number of magnets	---	120
θ_{PM}	PM circumferential angle	deg.	12
W_{PM}	PM width	mm	9
W_{stc}	SMC tooth circumferential width	mm	9
W_{sta}	SMC tooth axial width	mm	8
H_{str}	SMC tooth radial height	mm	10.5
N_c	Number of turns of winding	---	125
D_c	Diameter of copper wire	mm	1.25
g_1	Air gap length	mm	1.0

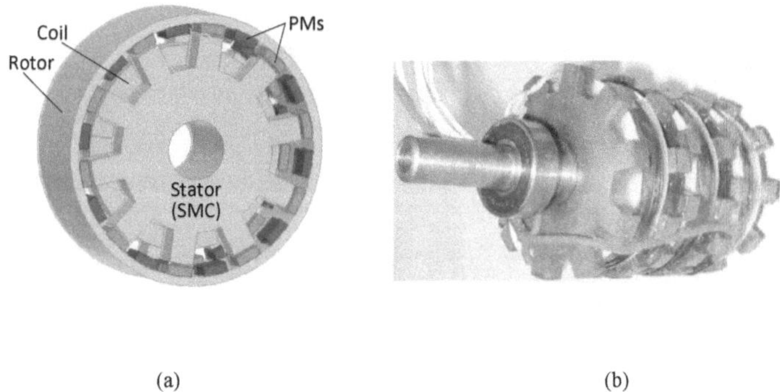

(a)   (b)

Figure 2-14 Structure of a PM-SMC TFM (a) and a prototype of the SMC stator (b).

For the multi-objective optimization model of (2-6), its robust design optimization can be written as

$$
\begin{aligned}
\min : \quad & \left\{ F_i\left[\mu_{fi}(\mathbf{x}), \sigma_{fi}(\mathbf{x}) \right], \quad i = 1, 2, \ldots, p \right\} \\
\text{s.t.} \quad & g_j\left[\mu_f(\mathbf{x}), \sigma_f(\mathbf{x}) \right] \leq 0, j = 1, 2, \ldots, m \\
& \mathbf{x}_l + n\sigma_{\mathbf{x}} \leq \mathbf{\mu}_{\mathbf{x}} \leq \mathbf{x}_{\mathbf{u}} - n\sigma_{\mathbf{x}} \\
& \mathrm{LSL} \leq \mu_f \pm n\sigma_f \leq \mathrm{USL}
\end{aligned}
\tag{2-7}
$$

To illustrate the advantages of this method, several examples have been investigated for different types of PM motors in terms of single and multi-objective situations [71, 72].

Results showed that the DFSS robust optimization method could provide design schemes with high reliabilities (lower probability of failure) and/or qualities (for example, smaller standard deviations for output power) for electrical machines. This is very valuable for engineering batch production.

This case study is based on a PM TFM with SMC cores. The robust analysis is based on the MCA and Kriging model of the motor. Figure 2-14 shows the structure of the motor and a prototype of the SMC stator. It was designed to deliver an output power of 640 W at 1800 rev/min in our previous work to validate the effectiveness of the application of SMC in motor design. Table 2-6 lists values for the main parameters of the motor prototype [87].

The traditional multi-objective optimization of this machine targets low material cost and high output power for a given volume. To show the significance of robust optimization, a third objective, manufacturing quality level (*n*) is introduced in this case study. The three-objective optimization model can be written as

$$\min : \begin{cases} f_1(\mathbf{x}) = \cos t \\ f_2(\mathbf{x}) = 1000 - P_{out} \\ f_3(\mathbf{x}) = 7 - n \end{cases}$$
$$\text{s.t.} \quad \begin{aligned} g_1(\mathbf{x}) &= 0.795 - \eta \le 0 \\ g_2(\mathbf{x}) &= 640 - P_{out} \le 0 \\ g_3(\mathbf{x}) &= sf - 0.8 \le 0 \\ g_4(\mathbf{x}) &= J_c - 6 \le 0 \end{aligned} \qquad (2\text{-}8)$$

where *sf* is the fill factor, η is the efficiency, P_{out} is the output power, and J_c is the current density of the winding. From previous design experience, four parameters (θ_{PM}, W_{PM}, N_c and D_c) are significant to the performance of this machine. Therefore, they are selected as optimization factors in this case study. The manufacturing tolerances of them are 0.05 deg, 0.05 mm, 0.5 turn and 0.01mm, respectively.

Figure 2-15 illustrates the comparison of a three-objective optimization problem with consideration of manufacturing quality (model (2-8) itself) and a two-objective optimization problem without consideration of manufacturing quality (remove the third objective from the model (2-8)). In the figure, red squares and black circles are the Pareto fronts of them respectively; blue and pink points are the projections of red squares in the 2D planes. As shown, some designs (red squares on the top) have high manufacturing quality (6σ quality), while others' qualities are low (less than 4σ), especially for those extremely optimized designs like those points with output power over 750 W. For a general balance of the objectives (similar weighting factors), the designs of moderate objectives (output power around 740 W and active material cost around $ 32) would be adopted in many situations. Thus, a critical cylinder is illustrated in the figure. (in this case) All designs in this cylinder have similar performances, so all can be regarded as good candidates. However, considering the manufacturing quality, some designs have low manufacturing quality (less than 4σ), and some have high manufacturing quality (6σ). Therefore, if the designer ignores this aspect, the design of low manufacturing quality may be picked. To avoid this situation, manufacturing quality should be investigated in the design optimization stage.

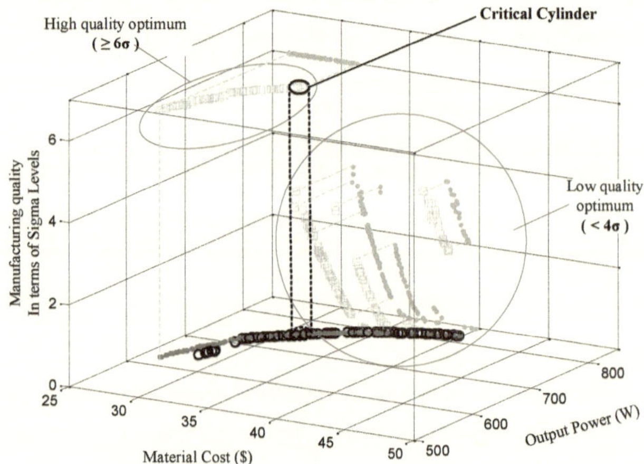

Figure 2-15 Illustration of design optimization for 6σ manufacturing quality.

2.10.3 A Case Study for Robust Optimization of Electrical Machines for High-Quality Manufacturing

This example aims to show the advantages of robust optimization by investigating a PM-SMC TFM. In this case, all eight parameters listed in Table 2-6 and the press size of the manufacturing the SMC core will be covered in a single objective optimization model [4]. The deterministic optimization model can be defined as,

$$
\begin{aligned}
\min: \quad & f(\mathbf{x}) = \frac{\cos t}{C_{m0}} + \frac{P_0}{P_{out}} \\
\text{s.t.} \quad & g_1(\mathbf{x}) = 0.795 - \eta \leq 0 \\
& g_2(\mathbf{x}) = 640 - P_{out} \leq 0 \\
& g_3(\mathbf{x}) = sf - 0.8 \leq 0 \\
& g_4(\mathbf{x}) = J_c - 6 \leq 0
\end{aligned}
\tag{2-9}
$$

where C_{m0} is the sum of the material cost and manufacturing cost of the SMC core. The robust optimization model of (2-9) can be expressed as

$$
\begin{aligned}
\min: \quad & \mu_{f(\mathbf{x})} \\
\text{s.t.} \quad & \mu_{g_i(\mathbf{x})} + n\sigma_{g_i(\mathbf{x})} \leq 0, i = 1, 2, \ldots, 4
\end{aligned}
\tag{2-10}
$$

Meanwhile, the probability of failure (POF) is employed as a criterion to compare the product's reliability of different designs given by deterministic and robust optimization methods.

$$
POF = 1 - \prod_{i=1}^{4} P\left(g_i \leq 0\right)
\tag{2-11}
$$

Table 2-7 lists several motor performances and the POFs for the deterministic and robust optimal designs. As shown, the cost and output power given by the robust optimal design are slightly worse than those of the deterministic optimum. However, after MCA with manufacturing tolerances, the POF of the deterministic optimum is 49.63%, while the POF of robust optimal design is nearly zero based on the given constraints. Through a further investigation, it is found that the high POF is mainly due to the constraint of winding current density. Figure 15 shows the MCA of the current density in winding for both optimums with a sample size of 10,000. As shown, the average is 6.00A/mm^2 for the deterministic design, which is exactly the same as the design limit. Consequentially, there are many samples exceeding this limit, which result in a big POF for this constraint. For the robust optimal design, all samples are below the limit, thus the POF of this constraint is 0. The lower cost and higher output power of deterministic optimum are obtained at the cost of high POF or low reliability and robustness. This is not acceptable from the perspective of industrial design and production.

Table 2-7 Comparison of the motor performances and the POFs

Par.	Cost	P_{out}	J_c	POF
Unit	$	W	A/mm^2	%
Deterministic	27.8	718	6.00	49.63
Robust	28.8	700	5.76	~0

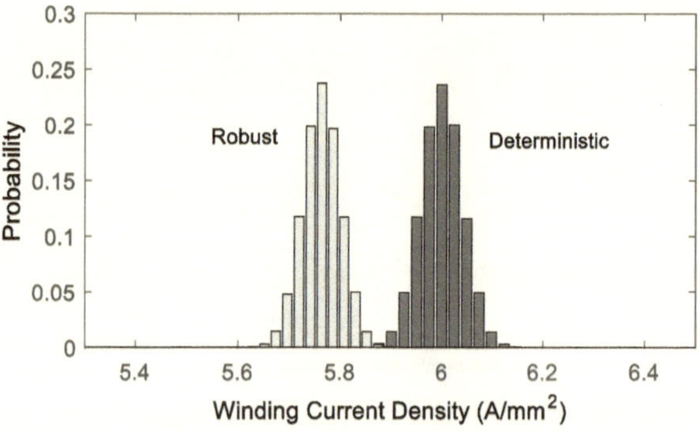

Figure 2-16 Comparison of current density between deterministic and robust optimizations.

2.10.4 Robust Design Optimization Methods with Non-Stochastic Uncertainties

The robustness measures by the stochastic based methods generally require the inherent properties of the uncertainties. To obtain the distribution of the uncertainties, huge amounts of samples are required. Therefore, in many cases, the a variation interval is more accessible than the specific distribution. Accordingly, the robust optimization methods for the non-stochastic uncertainties are developed. Similar to the robustness measure, the variation bounds of the objective performance and the worst case of the constraints are widely applied for the robustness measures.

Similar to MCA for stochastic uncertainties, the direct scan method for the worst-case estimation generally requires a huge amount of samples. The strategy of reducing the middle solution in the optimization process required to do the robustness measurements is also investigated in [88-90]. The worst-case and performance variation bounds are only scanned for the potential local, global optimum and solutions located on the boundary. A double loop optimization method is presented in [91] taking advantage of the Kriging model, in which the outer loop is for the optimal solution searching while the inner loop is for the worst-case searching.

The research on the application of the interval surrogate model such as Taylor extension is found for fast worst-case estimation [92-94]. However, besides the sensitivity calculation, the utilization of the Taylor extension method for the worst case may have a conservative solution because of the inaccurate evaluation of the bounds. In our research, the Chebyshev inclusion has been introduced for the interval uncertainty quantification, which shows an accurate worst-case estimation which is introduced in this chapter [95]. The application of the fuzzy set method for the interval uncertainty can be found in [96].

2.10.5 Comments

This section reviews the classification of uncertainties and techniques for uncertainty quantification and robust optimization. Two of the study cases present the importance of robust optimization for high-quality manufacturing. The uncertainty quantification for both cases is based on MCA and surrogate models. For efficient uncertainty quantification and robust optimization methods considering both stochastic and non-stochastic uncertainties, extending research on the robust optimization of electrical machines by the candidate is introduced in Chapter 6.

In addition to the robust optimization considering manufacturing uncertainties, the relative aspects include the manufacturing cost and manufacturing constraints, etc. are also of vital importance when we take the manufacturing conditions into consideration of the design optimization process of electrical machines. Some extending research on the combination of robust optimization with the manufacturing cost of different manufacturing accuracies and batches can be found in [97, 98]. The completion of the optimization considering manufacturing for electrical machines still requires great efforts on effective modeling of manufacturing conditions.

2.11 Application-Oriented System-level Design Optimization for Electrical Drive Systems

2.11.1 Method and Flowchart

As one of the components, the electrical machines generally compose the drive or generation system together with power and drive circuits, controllers, etc. The system-level performances such as system efficiency, cost, dynamic performance, etc. are more important compared with the single component performance of the electrical machine. The operation performances of electrical machines include not only the steady-state performances, such as average output power, average torque, and efficiency but also the dynamic responses, including speed overshoot, settling time and torque ripple. To estimate these steady-state performances, the multi-physics analysis of the motor is always required. To evaluate the dynamic responses, simulation analysis should be conducted for the control systems of the machines.

Meanwhile, more and more electrical machines and control systems are required to be integrated as one part (i.e. drive systems) in the applications, such as electric vehicles (EVs) and hybrid electric vehicles (HEVs). EVs and HEVs have attracted much attention from both government and industry due to the worldwide energy conservation and environment protection [99-101]. Electrical drive systems are crucial for the energy efficiency of the whole HEVs and EVs, which require integrated design and optimization, like the in-wheel motor drive. Therefore, design optimization of the whole electrical drive systems has become a promising research topic recently because assembling individually optimized components like a motor into a drive system cannot necessarily guarantee an optimal system performance [102]. For example, the efficiency maps and parameters of the electric machine, inverter, gearbox, battery, etc. on the component level are very important to reach the lowest overall power consumption of EVs or HEVs during several drive cycles of the car (system level). Partly low-efficiency points may be acceptable on the component level if another component compensates this to reach a high system performance.

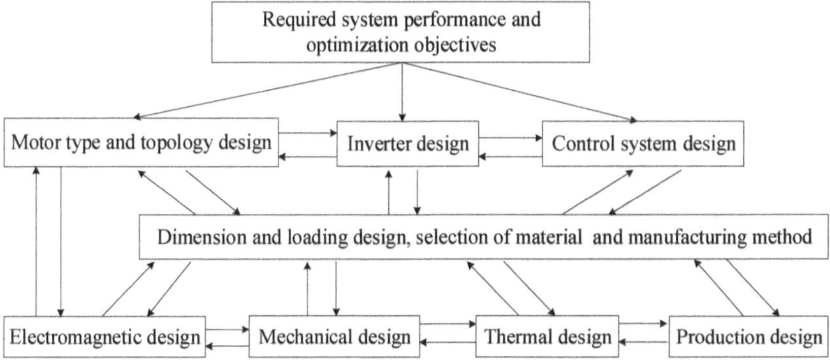

Figure 2-17 Framework of system-level design optimization method for electrical drive systems.

As above summarized, the application-oriented system-level design optimization of electrical machines and their drive systems can be regarded as the unification of the techniques for the system design optimization problem. The system-level performance consideration increases the freedom of the design optimization problem. For the system level consideration, from one aspect we should explore and include more design possibilities such as the new material application and performance modeling. However, the corresponding design optimization is a challenge as it is an application-oriented system-level problem instead of a component-level problem. It is also a multi-disciplinary, multilevel, multi-objective, high-dimensional and highly nonlinear design problem. Figure 2-17 illustrates a brief framework for system-level design optimization for electrical drive systems which gives consideration to the power electronics and control system design. Note that the drive system can be more complicated than the one in this figure if more components are considered. Compared with the optimization of electrical machines, the model is more comprehensive which aggravates the computation burden further. Effective optimization methods and strategies as introduced above such as the model reduction should still be applied to transfer the computationally prohibitive problem into one which can be solved.

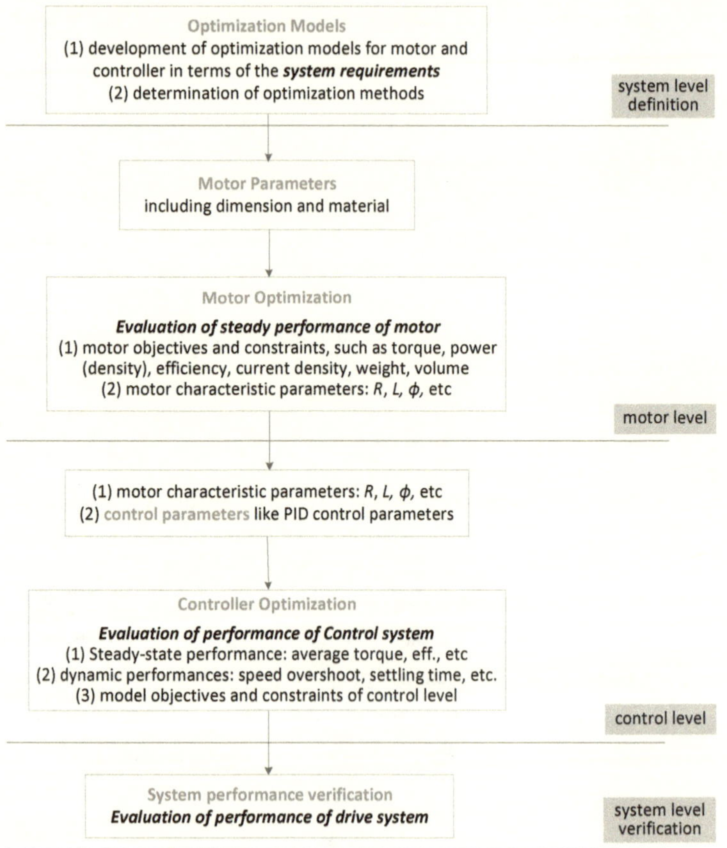

Figure 2-18 Multilevel optimization method for an electrical drive system.

To attempt this challenging problem, a multilevel optimization method as shown in Figure 2-18 has been developed for an electrical drive system with a specific motor and control system [72, 103]. As shown, there are four main steps and three levels (motor level, controller level, and system level) with effective coupling parameters between each other. The first step is to define the optimization models for motor and controller levels in terms of the system requirements. The second step is to optimize the motor. This level aims to optimize the motor performances in the steady-state, such as torque and efficiency. The third step is to optimize the controller performance. This level targets to optimize the dynamic performances of the drive systems based on specific control strategies and algorithms, such as field-oriented control (FOC), direct torque control (DTC), and model predictive control (MPC). Algorithm parameters such as factors of the proportion integration differentiation (PID) controller and duty cycle are the main optimization parameters at this level. The dynamic responses can be simulated by MATLAB SIMULINK. The last step is about system-level again. It aims to verify the system performances based on the obtained optimal design scheme regarding the design specifications. Particularly, considering the practical manufacturing and production specifications such as the manufacturing cost and robustness, the manufacturing technology and production design are to be considered in the optimization framework.

2.11.2 Comments and Suggestions

First, the design optimization problems are high dimensional, especially for the motor. To improve the optimization efficiency, a multilevel optimization method discussed in section 2.7.3 can be employed which coordinates with the system-level performance optimization.

Based on a case study of a drive system consisting of a PM-SMC TFM and MPC control [103], it was found that both the objectives for motor level and control level have been improved for this drive system by using the multilevel method. Better system steady-state and dynamic performances have been obtained. Meanwhile, the computation cost of the proposed optimization method is less than 20% of the motor level, and less than 50% of the control level, compared with those required by the direct optimization method. Therefore, the proposed method is efficient.

Second, the proposed method as shown in Figure 2-18 applies to a specific electrical drive system for an application. Normally, there are many options for a given application. For example, consider the motor drives in EVs, induction motors, permanent magnet synchronous motors (PMSMs) and switched reluctance motors are options for the motor part, FOC, DTC and MPC are choices for the control part. Moreover, each kind of motor has different types and topologies. Therefore, all of them should be optimized with a similar structure to obtain a fair conclusion.

Third, for the optimization objectives of the motor, some Pareto optimal solutions can be provided within the framework of multi-objective optimization. All of them should be evaluated for dynamic performances with a similar structure. Thus, the proposed method flowchart may be conducted several times in terms of different points in the Pareto solution.

Fourth, as optimization of electrical drive systems is a multi-disciplinary problem, the multi-disciplinary design optimization (MDO) methods developed in the field of aerospace and mechanical engineering can be employed for electrical machines. For the implementation, all disciplines are optimized through a parallel structure with a system-level optimization module which can coordinate the optimization between the main level and disciplinary levels. Though only a few discussions can be found in the literature for the MDO application on the electrical machines, this method presents another opportunity for efficient design optimization of electrical drive systems when more parameters, objectives, and disciplines are involved.

2.12 Conclusions

The design optimization methods of electrical machines are reviewed in the literature survey. The basic design optimization process is introduced. According to the general design optimization model, the relative components including the selection of design variables, the definition of objectives and constraints. Then the algorithms and techniques for effectively searching optimal solutions are reviewed in detail. For more automatic, high-freedom and integrated development processes of electrical machines and drive systems, three challenging topics are presented. They are topology optimization, robust optimization, and application-oriented system-level design optimization. This Book mainly focuses on the former two optimization methods.

Chapter 3 Electromagnetic Design of an In-wheel Motor with Grain-Oriented Silicon Steel

3.1 Introduction

Confronting the increasing of the requirement of energy, fossil fuel shortage and its contribution to environmental pollution, the new energy technology has been the hot research topic. On one hand, new energy source development has been widely investigated which includes marine energy, hydroelectric, wind, geothermal and solar power, etc. On the other hand, energy conservation technologies are also studied for reducing energy consumption. New energy vehicle is the integration product of the new energy production and consumption. Vehicles equipped with an internal combustion engine (ICE) have been in existence for over a hundred years. Although ICE vehicles are being improved by modern automotive electronics technology, they need a major change to significantly improve the fuel economy and reduce emissions [1]. EVs and HEVs have been identified to be the most viable solutions to fundamentally solve the problems associated with ICE vehicles.

Electrical drives are the core technology for EVs and HEVs. The basic characteristics of an electric drive for EVs mainly include high torque density and power density, a very wide speed range (covering low-speed crawling and high-speed cruising), high efficiency over wide torque and speed ranges, high reliability and robustness for the vehicular environment and acceptable cost. In addition, low acoustic noise and low torque ripple are important design considerations [2, 3].

Regarding the drive train topologies, most EVs are equipped with a centrally arranged electric motor that powers the front or rear wheels via a differential gear (DG) and a central controller which drives the inverter through switching signals (SSs), as shown in Figure 3-1 (a). One disadvantage of this concept is that it hardly reduces installation space in comparison to conventional vehicles due to the common arrangement of gears and driveshafts etc.[4]. Moreover, braking tests with single drive electric vehicles including measurements of recuperation energies and power have shown that a big amount of energy is still transformed into thermal energy by the friction brakes. Therefore, electric vehicles with independent drive motors including two-wheel-drive and four-wheel-drive promise better results.

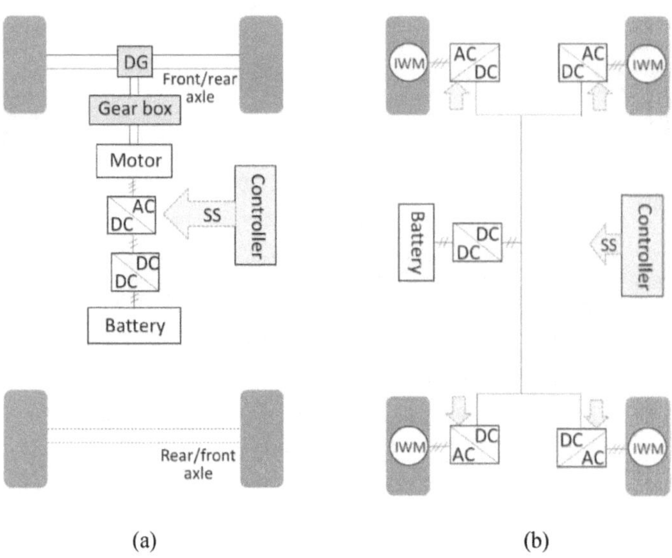

Figure 3-1 Typical drive train topologies for EVs, (a) single motor drive with one central motor, (b) four-wheel drive with four in-wheel motors.

In-wheel motor technology offers a solution to the distributed drive problem. An in-wheel motor is an electrical machine that is incorporated into the hub of a wheel and drives it directly as shown in Figure 3-1 (b). It replaces the transmission, driveshaft, axles, and differentials normally used in a conventional vehicle. Hence, in-wheel motor drive systems for EVs have the advantages of effectively maximizing interior space and providing improved vehicle driving performance due to the independent traction control ability of each wheel. The degree of freedom in vehicle design is increased greatly by an in-wheel motor system. In addition, with the removal of the differential and the driveshaft, energy loss is reduced and this allows for a reduction in electricity consumption required when running, thereby allowing the driving range per charge to be increased. Furthermore, the in-wheel motor directly drives the vehicle, so it can achieve highly responsive control of the driving force.

However, in-wheel motors have been around for more than a century, and the fact they haven't caught on speaks for themselves. To make in-wheel motors viable, there are some hurdles that need to be cleared. The main challenges include high unsprung weight and effective cooling (in a limited completely sealed space), etc.[5, 6]. Unsprung weight is the mass of all components not supported by a car's suspension. Conversely, sprung weight is the mass supported by the suspension, including the frame, motor, passengers, and body. Unsprung weight includes wheels, tires, and brakes, and it travels up and down over any bumps, potholes, and debris as it tries to follow the contours of the road. The sprung mass, however, is shielded from most of these movements, especially the smaller ones, by the suspension. And the sprung weight and suspension act to press down on the wheels so that they are in contact with the road. As a rule, designers try to minimize the unsprung weight to improve handling and steering. Adding in-wheel motors could vastly increase the unsprung weight and degrade performance. With in-wheel motors being part of the vehicle's unsprung weight, they will feel the impact of every pothole, bump, and high-speed turn.

For the problem of increased unsprung weight, efforts should be made on the improvement of both the motor design and suspension design, etc.On the aspect of the motor performance enhancement, this research aims to explore the design of a high-performance PM in-wheel motor. To meet the requirements of high torque/power density and high efficiency, a dual-rotor axial flux motor with grain-oriented material is developed. The specific electromagnetic analysis and design process are introduced in this chapter.

3.2 Design Requirements and Aims

The design space and operation speed of the in-wheel motor according to the requirements should be considered initially. As two widely used wheel hubs [7], the structure and dimensions of the wheel with 16-inch and 17-inch hubs are presented as Figure 3-1.

The rim diameter and width of the 16 and 17-inch hubs are 406mm, 165mm, 432mm, and 191mm. The rim width usually has various sizes for different section widths of the wheels. Meanwhile, the design space is also related to rim structure as well as inner mechanical parts such as the bearings, brakes, and supporting components.

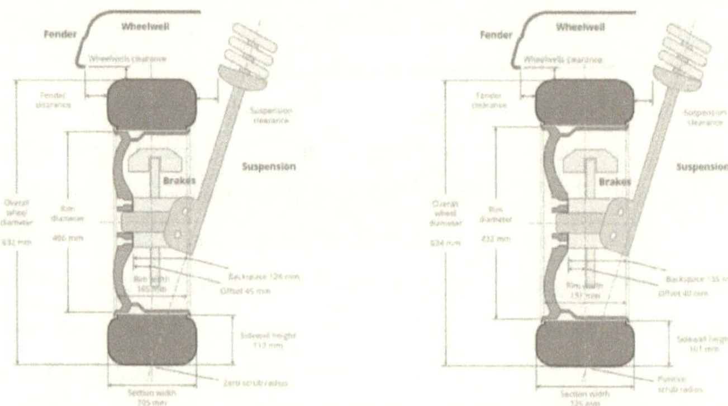

Figure 3-2 Structure and dimension parameter estimation of the 16 and 17-inch wheel hubs.

The operation capacity of the motor depends on the car drive and the overall wheel diameter. Given the overall wheel diameter is roughly 600mm, in the Australian context, the required speed range covers constant torque operation for 0-50 km/hr (0-450 rev/min), and constant power operation for 50-200 km/hr (450-1750 rev/min), with the maximum efficiency operation in the cruise speed range of 50-110 km/hr (450-1000 rev/min).

Paper [5] offers a 16-inch in-wheel motor design example in which the maximum external diameter and the maximum axial length are designed to be 362 and 105 mm, respectively, so that the proposed motor can be fitted into a 16-in aluminum wheel rim. The example presented in [5] basically matches the requirements. At the rated operating point (400rpm), the speed of a vehicle reaches 45 km/hr, at the operating point of 450 r/min, the speed of a vehicle reaches 50 km/h, and at of 1200 r/min, the speed reaches up to 136 km/h.

Table 3-1 Design specifications of the in-wheel motor

Performance	Unit	value
Rated power	kW	>18
Rated speed	rpm	400
Outer diameter	mm	<360
Axial length	mm	<130

Considering the above-mentioned design requirements on the motor dimension, speed, and output power. The design specifications of the motor are listed in Table 3-1.

3.3 Electromagnetic Design Analysis

According to the requirement introduced above, the design of the in-wheel motor is a tough question for a low weight design in oblate design space of high power/torque density and efficiency. Assuming the motor is designed with sinusoidal back EMF and excited with sinusoidal current with the same phase, the electromagnetic power can be expressed as

$$
\begin{aligned}
P_{em} &= \frac{m}{T}\int_0^T e(t)i(t)dt \\
&= \frac{m}{T}\int_0^T E_m \sin\left(\frac{2\pi}{T}t\right)I_m \sin\left(\frac{2\pi}{T}t\right)dt \\
&= \frac{m}{2}E_m I_m
\end{aligned}
\tag{3-1}
$$

where E_m and I_m are the amplitude value of the back EMF and current, and m is the phase number. Supposing the flux linkage ψ_m per turn is also pure sinusoidal

$$
\psi_m = -\psi_p \cos\left(\frac{2\pi}{T}t\right)
\tag{3-2}
$$

where ψ_p is the amplitude. The back EMF can be expressed as

$$
e = N_c \frac{d\psi_m}{dt} = N_c \psi_p \frac{2\pi}{T}\sin\left(\frac{2\pi}{T}t\right)
\tag{3-3}
$$

The electrical rotating speed is expressed as

$$
\omega_e = 2\pi / T = n_p \omega_r
\tag{3-4}
$$

Where n_p is the pole pair number, and ω_r is the mechanical rotating speed. The amplitude of the phase current is calculated as

$$
I_m = A_c J_m
\tag{3-5}
$$

The number of turns per phase is computed as

$$N_{c} = \frac{N_{s}}{m} \frac{K_{sf} A_{s}}{A_{c}}$$

(3-6)

Where N_{s} is the slot number, K_{sf} is the slot factor, A_{s} is the slot section area, A_{c} is the coil section area per turn. Then the electromagnetic power and torque can be expressed as

$$P_{em} = \frac{m}{2} N_{c} n_{p} \omega_{r} \psi_{p} I_{m}$$

(3-7)

$$T_{em} = \frac{P_{em}}{\omega_{r}} = \frac{m}{2} N_{c} n_{p} \psi_{p} I_{m}$$

(3-8)

or

$$T_{em} = \frac{N_{s}}{2} n_{p} \psi_{p} K_{sf} A_{s} J_{m}$$

(3-9)

It is obvious that if we want to achieve high torque, the phase number, pole number, flux linkage, and current can be considered. However, in the electrical machines of low operation speed with silicon steel core, the copper loss is usually dominant. That means the increase of the current aggravate the situation which leads to low efficiency and high-temperature rise. Therefore, we aim to enhance the torque density by improving the flux linkage and pole pairs in this case. As a result, the operation frequency and flux density will increase which balances the core loss and copper loss to reach high efficiency.

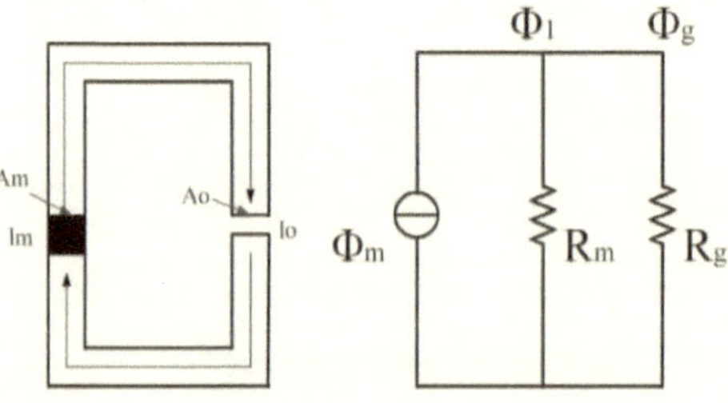

Figure 3-3 Magnetic circuit model.

Ignoring the reluctance of iron core and leakage flux, the flux that links the winding coil can be expressed as

$$\Phi_g = \Phi_m \bigg/ \left(1 + \frac{R_g}{R_m}\right) \tag{3-10}$$

where R_m and R_g are the reluctance of the PM and air gap respectively, Φ_m is the flux provided by the PM is calculated as

$$\Phi_m = B_m A_m \tag{3-11}$$

where B_m is the remanence and A_m is the cross-section area of the magnetic circuit.

To provide high flux linkage, for the same kind of PM with the same remanence, the flux depends on the section surface of the PM contact with the airgap.

As presented in Figure 3-4, there are three basic structures for the in-wheel PMSM design according to the rotor position with PMs, which are radial flux motors with the inner rotor and outer rotor, and dual rotor axial flux motor. For achieving the high flux linkage, we compare the section surface space that the three structures can offer. Obviously, the outer rotor design offers a larger section area for the PM than the inner rotor design. Here, we mainly compare the outer rotor design and dual rotor axial flux design.

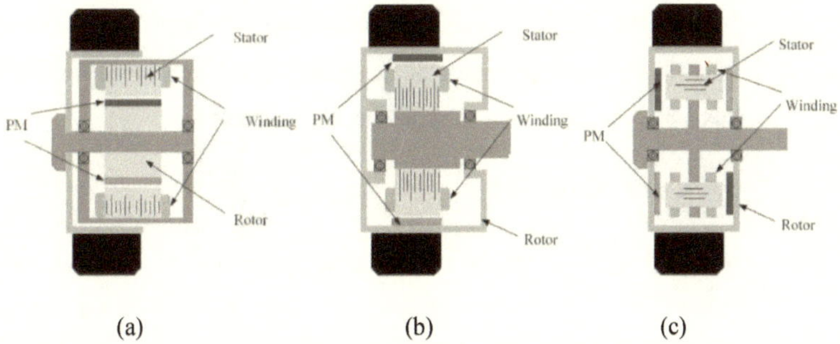

Figure 3-4 Three basic structures of the in-wheel PMSM (a) radial flux motors with inner rotor, (a) radial flux motors with outer rotor, and (c) dual rotor axial flux motor.

The area of side face used in the outer rotor motor for placing the PMs can be estimated as

$$A_{outer} = \pi D_{ext} L \tag{3-12}$$

where D_{ext} is the external diameter of the design space, L is the axial length.

The area of the end face of the axial flux motor with dual rotor for placing the PMs can be estimated as

$$A_{axial} = \frac{\pi}{2}\left(D_{ext}^2 - D_{in}^2\right) \tag{3-13}$$

where D_{in} is the inner diameter of the rotor. According to the maximum size for the design space, we assume the external diameter, the inner diameter of the rotor, and axial length are 362mm, 200mm, and 105mm [5]. Then the A_{outer} and A_{axial} can be calculated as 119412mm^2 and 143012 mm^2. Therefore, compared with the outer rotor motor, the axial flux motor has comparable and even more space for placing the PM in the design space to obtain higher flux. Furthermore, if embedded flux gathering structure as shown in Figure 3-7 is applied to the axial flux motor with dual rotor

$$A_{fg} = 4n_p l_{ar}\left(D_{ext} - D_{in}\right) \tag{3-14}$$

where l_{ar} is the axial length of the rotors. With the same aforementioned inner and outer diameter, the A_{axial} of a dual rotor design with 20 pole pairs and 12mm rotor length is 155520 mm^2 which means that the flux gathering structure can be used for increasing the flux linkage further for the dual rotor axial flux motor.

3.4 Core Material

As analyzed above, the dual rotor axial flux motor shows great potential for the application of in-wheel drive. Furthermore, the magnetic circuit of the dual rotor axial flux motor in the iron core shows the one-dimensional feature, i.e. flux path is almost in one axis. This characteristic provides the possibility of the performance-enhancing design with the anisotropic core material, i.e. grain-oriented silicon steel [8].

Table 3-2 Grain-oriented silicon steel sheets of various grades

Grade	Density (kg/dm^3)	Maximum Core loss (W/kg) at 1.7T		Minimum Induction (T)	Lamination factor
		50HZ	60HZ		
35ZH115		1.15	1.52	1.88	
35ZH125	7.65	1.25	1.65	1.88	0.96
35ZH135		1.35	1.78	1.88	

Table 3-3 Non-grain oriented silicon steel sheets of various grades

Grade	Density (kg/dm^3)	Maximum Core loss (W/kg) at 1.5T	Minimum Induction (T)	Lamination factor
		50HZ		
35HX230	7.65	2.30	1.66	
35HX250	7.65	2.50	1.67	0.95
35HX300	7.70	3.00	1.69	

Table 3-2 and Table 3-3 list the properties of the several grain-oriented silicon steel sheets and nongrain oriented silicon steel sheets of various grades but the same thickness (0.35mm) produced by Nippon Steel Corporation[9]. Compared with widely used nongrain oriented silicon steel sheet, grain-oriented silicon steel possesses higher magnetic permeability, higher saturation magnetic flux density, lower core loss, and lower magnetostriction, etc. However, the high magnetic conductivity is anisotropic, i.e. exists in only one direction [10]. This is also the reason that it is rarely applied as the core material in general electrical machines with 2 dimensional or even 3-dimensional flux path. Therefore, the application of the material may help gain more room for the improvement of motor performance. For the initial design, grain-oriented silicon steel 27ZH100 (0.27mm) is adopted as the motor core material. Figure 3-5 presents the induction and permeability variation of the 27ZH100 with respect to the magnetizing force in the rolling direction.

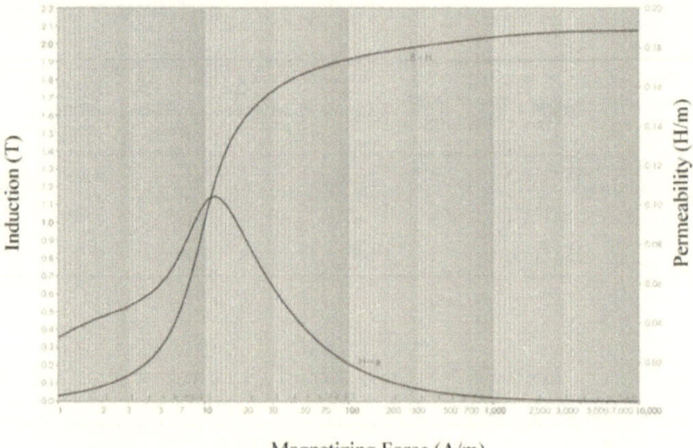

Figure 3-5 BH and permeability curves of the 27ZH100.

As presented in Figure 3-6, the core loss coefficients obtained by the curve fitting(W/m^3):

$$P_{coreloss} = P_{h} + P_{c} + P_{e} = k_{h}f\left(B_{m}\right)^{2} + k_{c}\left(fB_{m}\right)^{2} + k_{e}\left(fB_{m}\right)^{1.5} \qquad (3\text{-}15)$$

where the coefficients can be obtained by $k_{h} = 38.94$, $k_{c} = 0.253$, $k_{e} = 1.776$

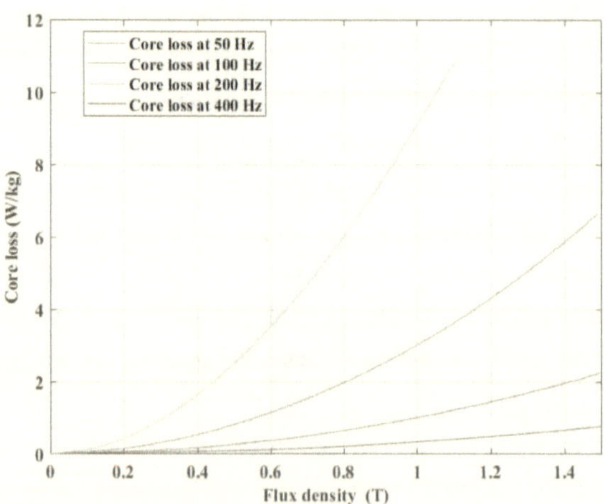

Figure 3-6 Core loss versus variable frequencies.

3.5 Motor Design

3.5.1 Winding Arrangement

Single-layer winding and double-layer winding are two general winding types. Single-layer winding is wound alternatively on the teeth, and each tooth is wounded a coil component when the double layer winding is selective. Qualitatively compared with single-layer winding, the double layer winding allows shorter end winding, lower harmonics of the magnetomotive force, lower torque ripple and eddy current loss in the PMs, more combinations of slots and poles. The back EMF of the double layer winding is more sinusoidal while the single-layer winding is more trapezoidal. The radical length of the design with double layer winding can be reduced compared with the single-layer winding. On the other hand, the single-layer winding possesses higher fault-tolerant ability because of the isolated winding component between the different phases [11, 12]. Considering the advantages of the double-layer winding mentioned above, the possible pole and slot combination of the design with 36 slots is explored.

Table 3-4 lists the fundamental winding factors for different pole numbers of the double-layer winding with 36 slots calculated by

$$k_{wv} = k_d k_p k_{sk} \tag{3-16}$$

The combination of the 36 slots and 42 poles is selected for achieving high pole number and then the high output power ability with the double layer winding design.

Table 3-4 Fundamental winding factors for different pole numbers of double-layer windings with 36 slots

Poles	24	26	28	30	32	34	38	40	42
k_{w1}	0.866	0867	0.902	0.933	0.945	0.953	0.945	0.953	0.933

Table 3-5 Design parameters of the in-wheel motor

Par.	Symbol	Unit	Value
Stator inner radius	Lsi	mm	109
Radial length of stator teeth	Lso	mm	51
Rotor inner radius	Lri	mm	109
Radial length of rotor teeth	Lro	mm	51
Axial length of stator teeth	Hst	mm	80.8
Inner width of stator teeth	Wsti	deg	5
Outer width of stator teeth	Wsto	deg	6.4
Axial length of rotor teeth	Hrt	mm	13
Lower inner width of rotor teeth	Wrti	deg	6
Lower Outer width of rotor teeth	Wrto	deg	5
Upper inner width of rotor teeth	Wrti1	deg	0.8
Airgap length	Hgap	mm	2.1
Number of poles	NP		42
Number of stator teeth	NS		36
Number of coils of each slot			28
Width of the copper wire section		mm	2
Length of the copper wire section		mm	4
Relative permeability of PM			1.4

3.5.2 Motor Structure and Design Parameters

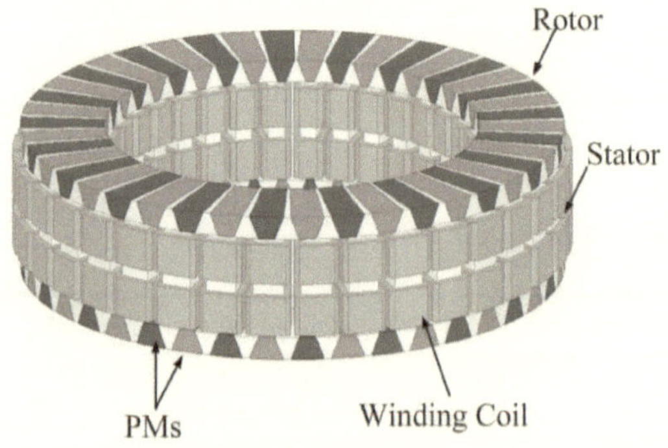

(a)

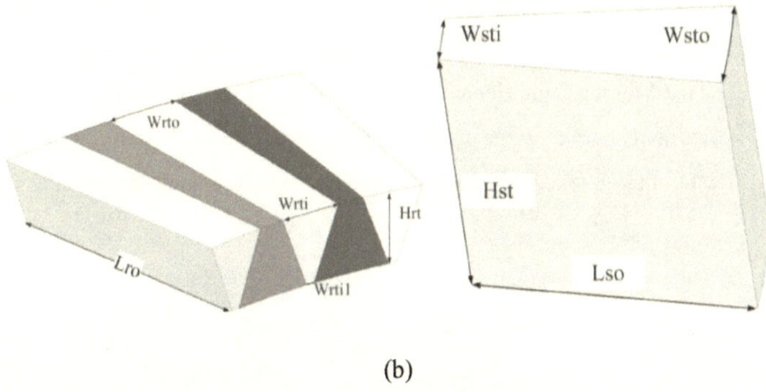

(b)

Figure 3-7 Motor structure (a) and topology parameters (b).

The motor structure and design parameters are presented in Figure 3-7. The middle part of the stator tooth without the winding coil is saved for the stator fixation. The introduction of the design parameters is listed in

Table 3-5 and the mass of the material is listed in Table 3-6. It should be noted that a large air gap length is designed here while a smaller air gap length can help increase the flux further. However, due to the limit current manufacturing condition, a smaller air gap cannot be realized. Thanks to the axial dual rotor design, the manufacturing difficulty comes from the requirement of the precision for preventing the unbalanced axial force of the stator caused by the uneven axial air gaps. The design with a smaller air gap will be discussed in the optimization section in the next chapter.

Table 3-6 Material mass of the initial design

Material	PM	Copper	Silicon steel	Total
Mass (kg)	5.1	11.1	18.5	34.7

3.6 Performance Calculation with FEM

Due to the complicated structure, 3D numerical techniques are required for accurately computing the motor parameters. Here, 3D magnetic field finite element analysis is performed to calculate the key parameters, such as the winding flux, back EMF, cogging torque, inductance, losses, maximum output torque and efficiency map in the operation range. Thanks to the symmetrical structure of the motor, the half model of the motor can be used for the finite element analysis with the symmetrical boundary condition, in which the flux is normal to the symmetrical surface. The mesh graph of the FEM model can be found in Figure 3-8 (a). The calculated performances are presented as follows.

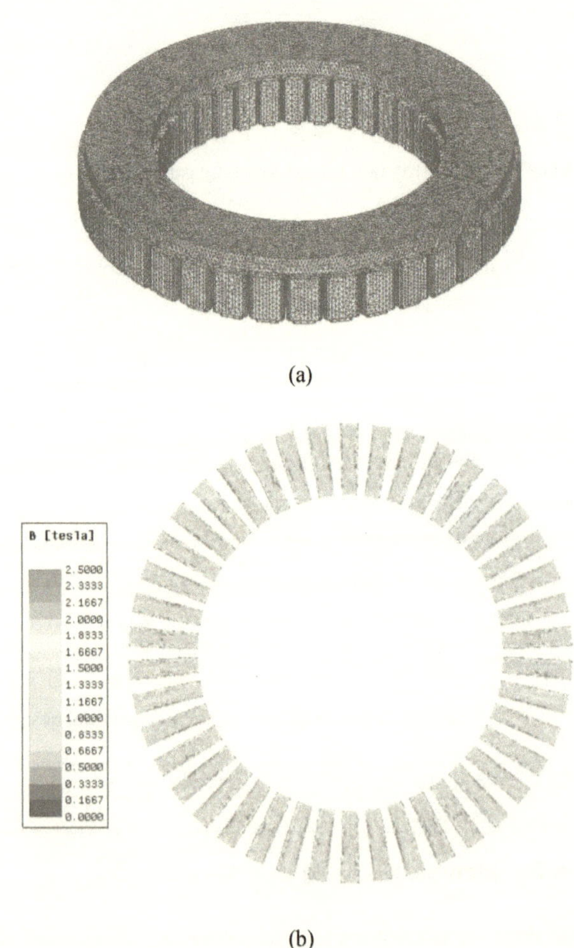

(a)

(b)

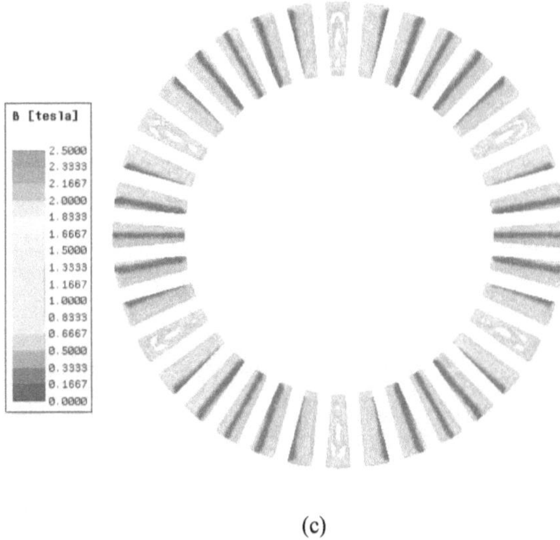

(c)

Figure 3-8 The mesh graph of the FEM model (a), and no-load magnetic field distribution of (b) rotor and (c) stator.

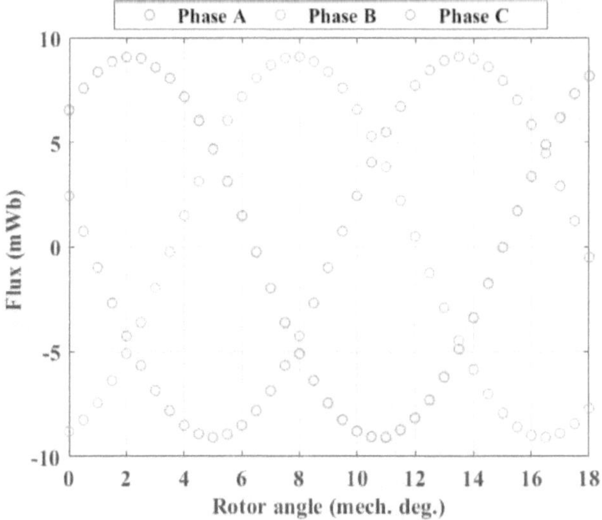

Figure 3-9 Per turn no-load flux of a phase winding.

3.6.1 Flux Linkage and Back EMF

The PM flux, defined as the flux of a stator phase winding produced by the rotor PMs, can be obtained from the no-load magnetic field distribution Figure 3-8. The flux waveform can be calculated by rotating the rotor magnets for one pole pitch in several steps, as shown in Figure 3-9. The effective value of back EMF of a single-phase at rated speed 400rpm can be calculated as

$$E = 4.44\,fN\Phi = 157.9\text{V} \qquad (3\text{-}17)$$

where $f = 140$ Hz, N is the number of turns of a tooth is which is 28 and Φ is the magnitude of the flux waveform of one phase with one turn coil which is computed as 0.00907Wb.

3.6.2 Resistance and Inductance Calculation

The stator winding resistance can be calculated by

$$R_1 = \frac{\rho_t l_1}{A_1} \qquad (3\text{-}18)$$

where l_1 is the total wire length in m, A_1 the wire cross-section area in m^2, and ρ_t the electrical resistivity in Ωm of the stator winding at temperature t which can be calculated as

$$\rho_t = \rho_0(1+\alpha t) \qquad (3\text{-}19)$$

where ρ_0 is the resistivity at 0 centigrade equals to 0.165 Ωm, α is the temperature coefficient equals to 0.0039. The average length of each turn of the coil is 0.149 m, the total length of the phase winding considering the end winding and the outgoing line is 55.6m. the cross-section area of the copper wire after the rounding is $7.637 \times 10^{-6}\,\text{m}^2$. The phase winding resistance can be obtained as 0.129Ω in the room temperature of 25°C. The computed self and mutual inductances at different rotor angles are presented in Figure 3-10.

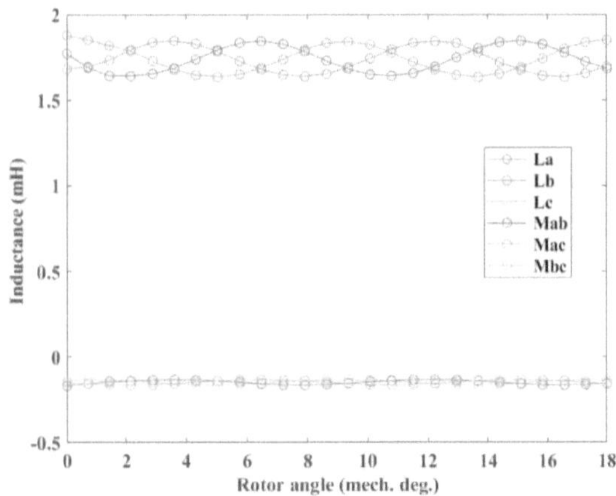

Figure 3-10 Computed self and mutual inductance of three-phase winding.

3.6.3 Cogging torque

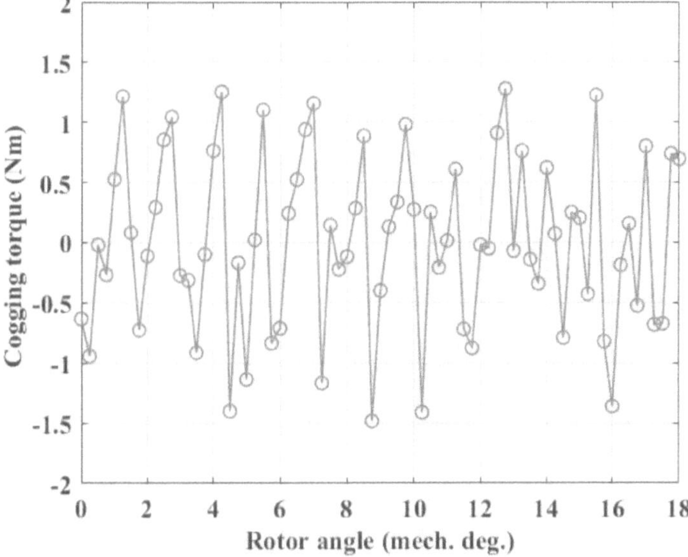

Figure 3-11 Cogging torque.

Cogging torque is caused by the tendency of the rotor to line up with the stator in a particular direction where the magnetic circuit has the highest permeance. The cogging torque arises from the reluctance variation of the magnetic circuit as the rotor rotates and exists even when there is no stator current. Figure 3-11 presents the cogging torque variation with the rotor angle. For the current design, the cogging torque can be negligible compared with the output torque. The simulated no load core loss is presented in Figure 3-12

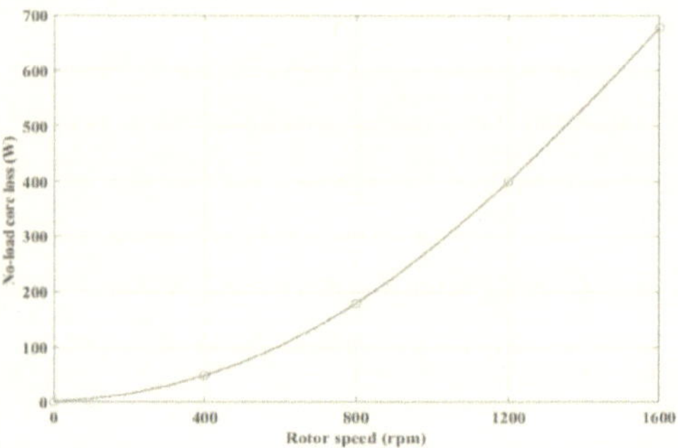

Figure 3-12 No load core loss.

3.6.4 Steady-State Performance Prediction

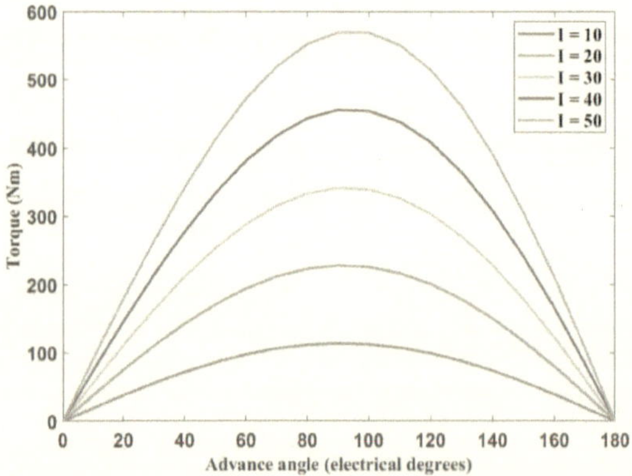

Figure 3-13 Simulated torque vs. advance angle curves of various currents.

The output torque, power, input power, and efficiency can be calculated as

$$T_{\text{out}} = P_{\text{out}} / \omega_r \tag{3-20}$$

$$P_{\text{in}} = P_{\text{out}} + P_{\text{cu}} + P_{\text{fe}} \tag{3-21}$$

$$P_{\text{cu}} = 3I_1^2 R_1 \tag{3-22}$$

$$\eta = P_{\text{out}} / P_{\text{in}} \tag{3-23}$$

According to the calculation results of the FEM, the output torque vs. advance angle curves at various currents are demonstrated in Figure 3-13. At the rated speed 400 rpm, the maximum output torque of the rated 40 A phase current is 450 Nm. The calculated copper loss and iron loss are 604 W and 77 W, respectively. The efficiency at the rated operation point is 96.6%, and the output power is 19 kW.

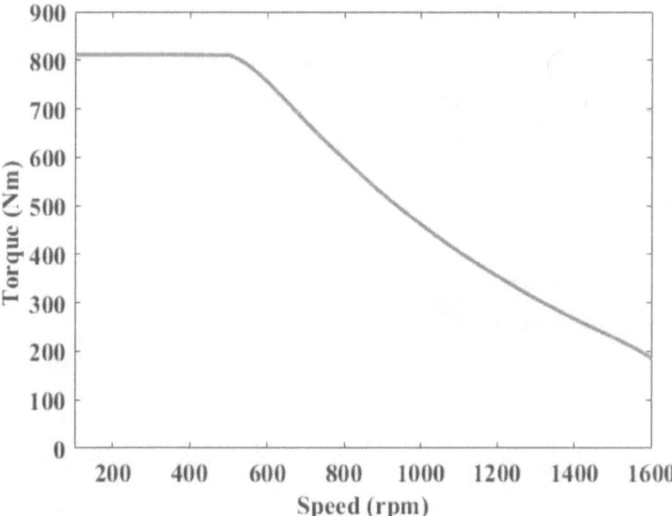

Figure 3-14 Maximum torque envelope.

Powered by the inverter of 600 V DC voltage with 70 A phase current limit, the maximum torque envelope is computed and presented in Figure 3-14. The current design meets the design specifications on the rated power while the maximum torque is more than 300Nm at the rated speed. The core loss, inductance, copper loss can also be regarded as a function with respect to the phase current, advance angle, and speed. Then we sampled enough FEM points to build the surrogate models of performance to those variables. Then for the operation points enveloped by the maximum torque envelope, we adopt the maximum torque per ampere (MTPA) control method to determine the minimum phase current and the advance angle at each operation point of specific speed and torque. Then we substitute the phase current, advance angle, and speed to the surrogate models to calculate the core loss, copper loss and efficiency map. The efficiency map is presented in Figure 3-15. The loss distributions including the copper loss, iron loss, and PM loss are demonstrated in Figure 3-16. The copper loss accounts for the main parts in the whole operation range, while the PM loss increases with the speed and raised maximumly to about half of the copper loss in the high-speed deep flux weakening region. Due to the high performance of the grain-oriented silicon steel, the core loss are the lowest among the three kinds of losses. Even in the high-speed region, the core loss is roughly one in sixth of the copper loss, and one in third of the PM loss, and the relatively low loss core loss contributes to the wide range of the high efficiency.

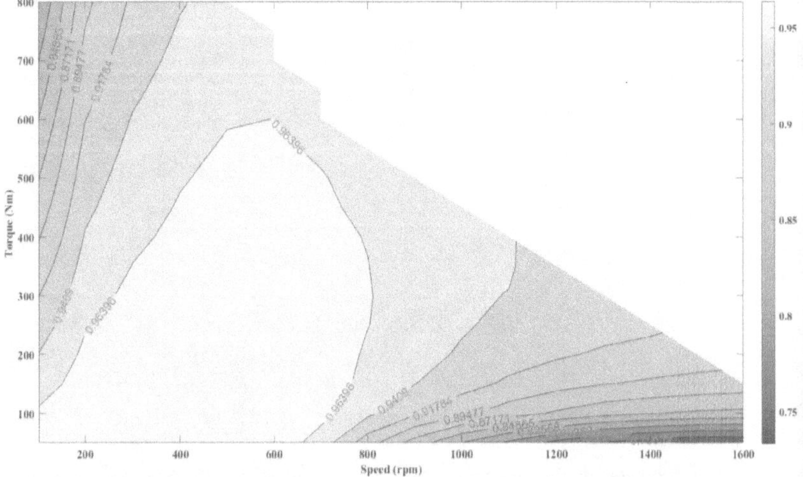

Figure 3-15 Efficiency map.

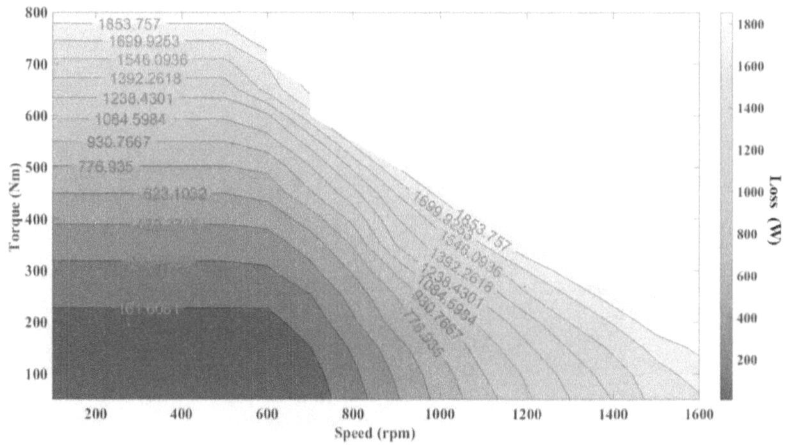

(a)

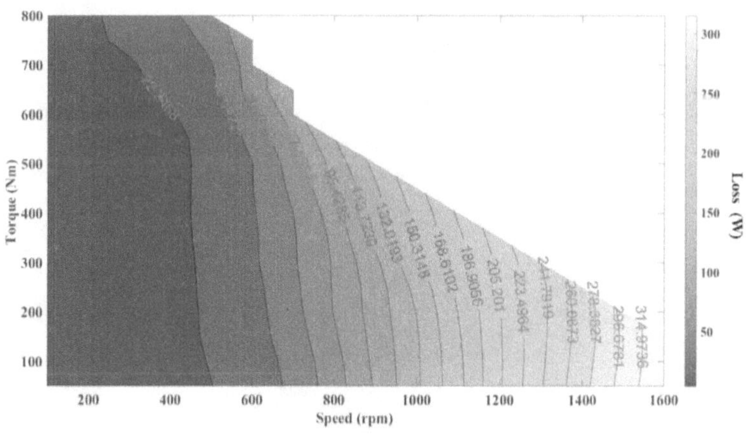

(b)

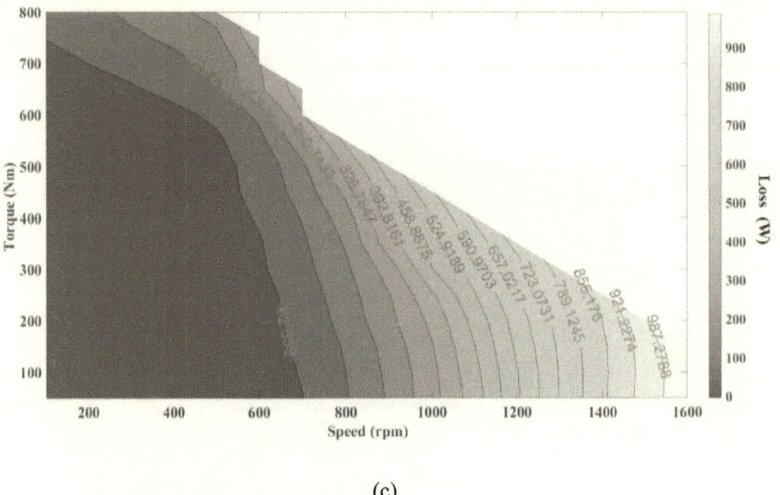

(c)

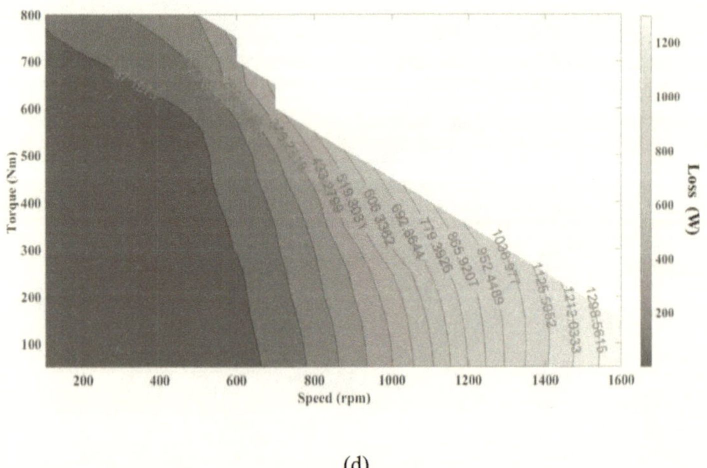

(d)

Figure 3-16 Map of the (a) copper loss, (b) core loss, (c) PM loss, (d) PM loss and core loss.

3.7 Conclusions

After the electromagnetic performance analysis, design and model verification, the current design of the dual rotor axial flux motor shows promising performance in output torque, efficiency while meets the basic design specifications. Due to the restriction of the current manufacturing condition in realizing the narrower air gap, effective coil winding, etc, the torque/mass ratio, i.e. the torque density of the current design is limited. By the investigation of the manufacturing technique, and motor test, the optimization of the mechanical and electromagnetic design is investigated in the following chapter.

Chapter 4 Prototyping, Test, and Optimization of the In-wheel Motor

4.1 Introduction

The manufacturing process of the designed motor topology is presented in this chapter which including the main process of the component manufacturing and assembling. Subsequently, the tests of the motor performance are described in detail. On one hand, the prototyping and the performance test aim to verify the accuracy of the design analysis model. As the key steps, we hope to accumulate the practical manufacturing experience by the motor prototyping for promising the desired electromagnetic performance. On the other hand, according to the test results, verification and modification can be conducted for the initial design model, which offers the firm basis for the subsequent optimization process.

4.2 Accessory Manufacturing

4.2.1 Stator Core Component

Since both stator and rotor core are radically laminated with grain-oriented silicon steel, two manufacturing schemes are proposed initially. The first approach is to laminate the steel sheet manufactured by punching. Since the lamination of the sheet are along the radical direction which means the size of every single sheet is different. This means a series of punching model should be manufactured. This approach may be with high efficiency at the batch production. However, the high economic and time cost for the series of model developing is not acceptable for the experimental stage. Another method for steel sheet manufacturing is laser cutting or wire-electrode cutting. This method is more flexible and cheaper for a small amount of production. Due to the limitation of lack of laser cutting machines, the wire-electrode cutting is selected for the sheet cutting. Based on this cutting condition, the fixation of laminations is another problem should be considered. To simplify the fixing process, the preprocessing of the sheets by punching the same size steel sheet with slots for fixation is conducted. Then the sheets can be laminated as the tooth segment for cutting which increases the cutting efficiency. Particularly, mortise and tenon joints are designed which are used for fixing the stack. The stator and rotor tooth component can be manufactured by similar steps. Figure 4-1 and Figure 4-2 present the mold designs for the silicon steel sheet punching. Figure 4-3 illustrates the punching process. The sheets after the manufacturing are presented in Figure 4-4.

Figure 4-1 Mold of the stator sheet.

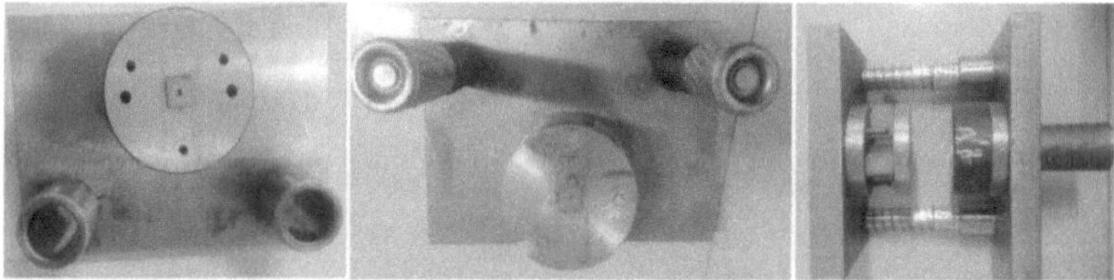

Figure 4-2 Mold of the rotor sheet.

Figure 4-3 Punching process of the silicon steel sheet.

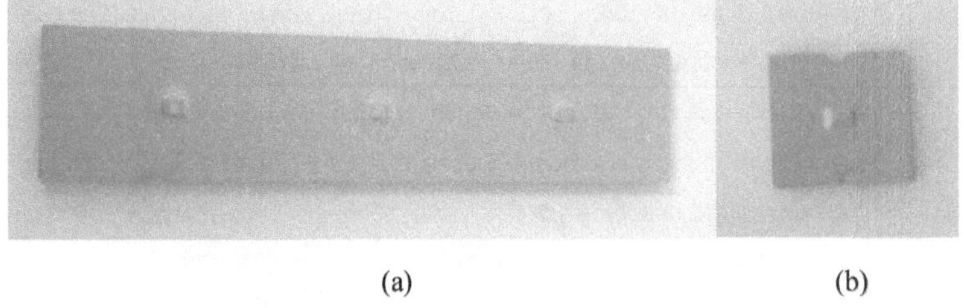

(a)   (b)

Figure 4-4 Punched (a) stator sheet and (b) rotor sheet.

To accelerate the lamination process while keep the stack quality with a constant lamination factor, stacking molds are manufactured as demonstrated in Figure 4-5. Figure 4-6 shows the tightly mounted stack fabricated with the stacking mold and pressing machine.

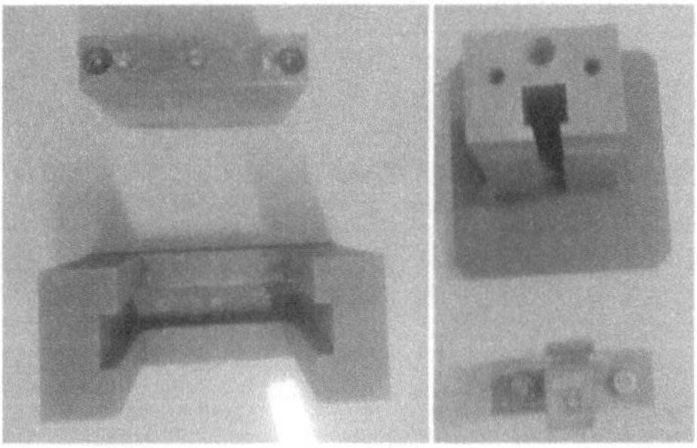

Figure 4-5 Stacking molds for stator and rotor cores.

Figure 4-6 Stator core stacking by the mold and stacked core.

As mentioned above, the lamination stack still requires the wire-electrode cutting to get the designed topology. For the requirement of fixation, steps in the middle axial of each tooth are machined for matching the slots in the fixing plate. Moreover, chamfering is also indispensable for winding and assembling. Figure 4-7 demonstrates the final shape of the segments.

Figure 4-7 Stator and rotor core after wire-electrode cutting.

4.2.2 Fixing Plate for Stator

The outer-rotor design of the motor is to meet the requirement of the in-wheel drive, in which the shaft of the motor is fixed while the transmission is through the rotor directly. Therefore, the stator part of the motor should be fixed on the motionless shaft, which differs from the machines with shaft transmission. Figure 4-8 exhibits the fixation plate of stainless steel in which trapezoid slots are manufactured by cutting. After fitting into slots, the stator components are then fixed by screws and blocks for pinching.

(a)

(b)

Figure 4-8 Stator tooth fixation plate, (a) side view and (b) front view.

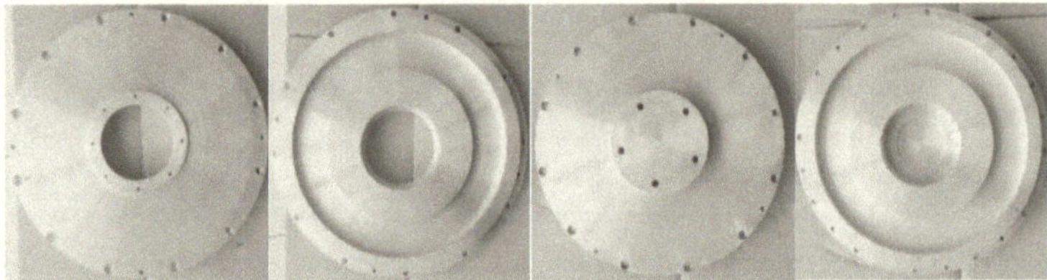

Figure 4-9 Aluminum alloy rotor carrier.

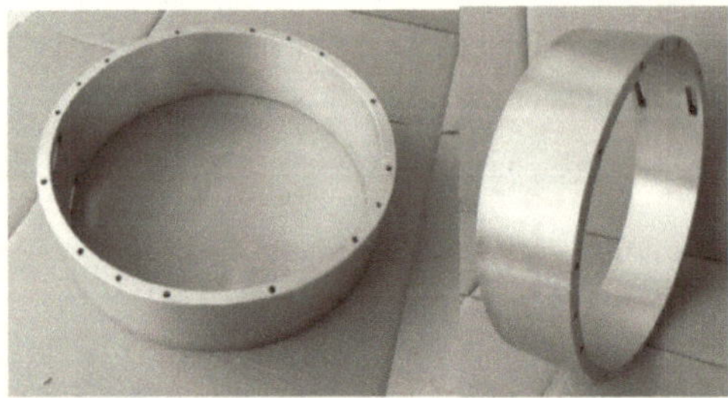

Figure 4-10 Connection hub.

4.2.3 Rotor Carrier and Connection Hub

Aluminum alloy is chosen for the housing of the rotor due to the property of the high mechanical strength, thermal conductivity, and low permeability. As illustrated in Figure 4-9,while sharing the same inner channel for placing the PMs and silicon steel sheet stack, the outside structure of the two aluminum rotor carriers differs from each other. One carrier with the inner hole is for the motor fixation by the shaft while the other is designed for the connection with the load machine through the flange. The hub connection part as shown in Figure 4-10, is designed for torque transmission between the two rotors since the output torque can only be transferred and tested from one side of the rotor. On the other hand, the connection part is designed to meet the sealing requirement for the in-wheel motor design. However, grooves are milled on the cylindrical surface of the connection ring for the convenience of the measurement in the assembly and testing processes, such as the air gap and stator temperature which are sealed during the operation of the motor.

4.2.4 Shaft

Due to the outer rotor structure of the axial motor, the through-hole is required for the winding wire threading through the shaft. Meanwhile, the shaft is also for supporting the stator and rotor which needs high mechanical strength. Therefore, hardening and tempering were firstly carried out for enhancing the strength, tenacity, and machinability of the steel material. Then the raw material is processed by machining, drilling, forming. For the coordination with the fixing plate, the shaft was then manufactured by threading and accurate grinding. Finally, the key grooves and holes on the cylindrical surface are milled for the fixation between the plate and shaft and drawing out the winding wires. Figure 4-11 exhibits the manufactured shaft.

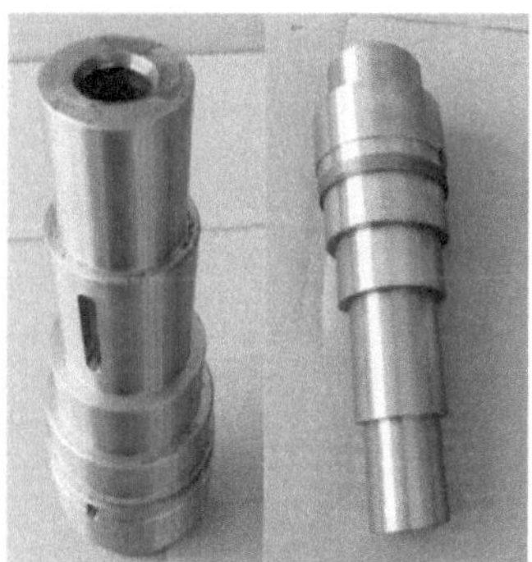

Figure 4-11 Shaft.

4.2.5 Winding Component and Other Components

Due to the application of flat copper wire and high slot fill ratio, the manufacturing of the winding component requires the professional winding machine. The insulation of the copper wires should be protected from being destroyed during the process such as twisting, while swell of the wires at the twisting parts should be avoided or decreased to an acceptable height in order to avoid the axial length of the component being higher than the slot. Figure 4-12 presents the winding components manufactured by the winding machine. As shown in Figure 4-13 - Figure 4-15, the other components including radial and axial bearings and PMs are ordered from the manufacturers directly.

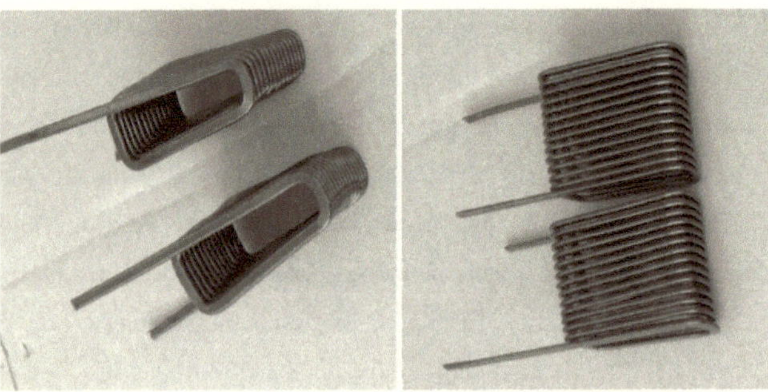

Figure 4-12 Winding component.

Figure 4-13 Radial bearing.

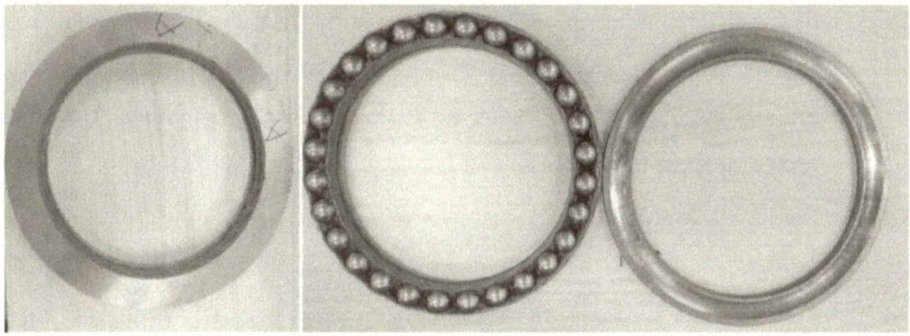

Figure 4-14 Axial bearing.

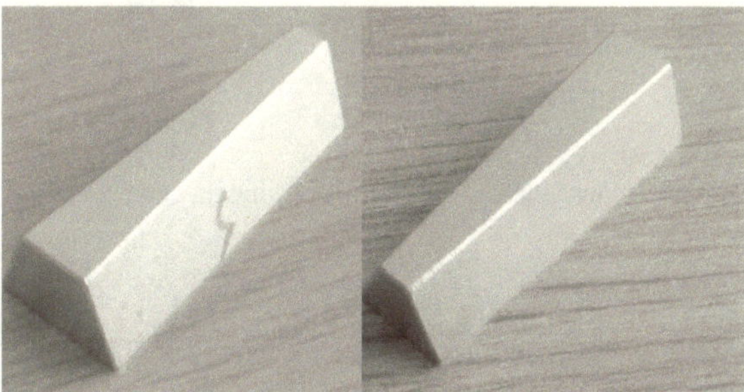

Figure 4-15 Permanent magnet.

4.3 Assembly

4.3.1 Stator Assembly

The winding component was firstly processed by dipping insulating heat conduction glue and forming. As shown in Figure 4-16, insulation paper is utilized for wrapping up the stator, winding coils and fixing plate for increasing the insulation while protecting insulation of the coils from abrasion. After the winding components are placed into the slots, the insulation between the fixing plate and the coils should be tested by the high voltage insulation tester. After the insulation performance is promised, insulating heat-conducting glue was then injected between coils and stator teeth to fix the winding components and increase the heat conduction performance. Then the winding components are connected according to the configuration by welding. The sectional area welding spot should be larger than the wire's while the destruction of the insulation of the wire should be avoided. After all the welding spots are finished, the joints should be processed by further insulating. The fixation between the stator and shaft is carried out by keys and grooves. As illustrated in Figure 4-17, after the shaft assembling, the winding ports are drawn out from the center hole of the shaft.

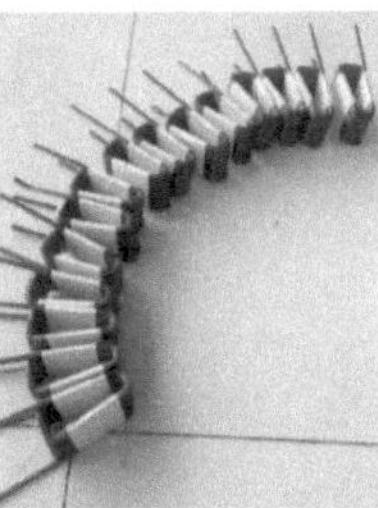

Figure 4-16 Insulation of the stator and winding components with insulation paper.

Figure 4-17 Stator assembly.

4.3.2 Rotor Assembly

The assembly of the rotor mainly contains the work of putting the PMs and silicon steel components into the slot of the aluminum carrier and gluing them together with industrial adhesives. The PMs are adhered to the carrier with the guidance of location holes processed by the machining center. As shown in Figure 4-18 (a), the silicon steel stacks are started to be glued after each PM is fixed. On contrary to the method of gluing the PM and silicon steel stack alternatively, the tolerance accumulation can be effectively eliminated in this way.

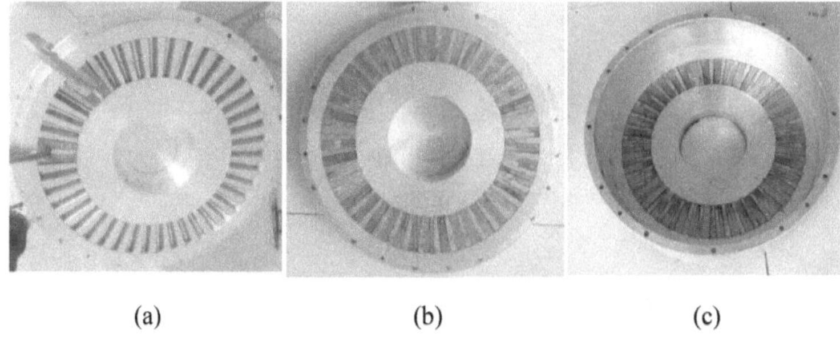

(a)   (b)   (c)

Figure 4-18 Rotor assembly.

4.3.3 Final Assembly

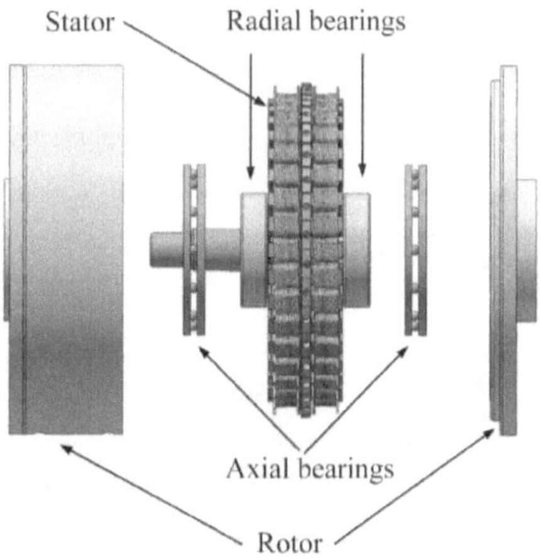

Figure 4-19 Final assembly.

Due to the multiple-pole and magnetic flux gathering design, the axial force between the rotor and stator is very high. In the ideal condition, the airgaps between In the stator and rotors can be kept equal during the assembly process and the axial forces of the stator are balanced. Otherwise, there is the risk that the stator and rotor are drawn together during the assembly. However, it requires the axial lengths of the stator teeth at each side of the fixing plate are equal, while the rotors can be assembled simultaneously with the same speed and relative distance to the stator. Furthermore, fixing strength of the stator and rotor should be guaranteed in case of the uneven of the air gaps during motor operation because of vibration. Due to the existence of tolerances and lack of automatic assembling equipment, the ideal condition can be hardly achieved. To reduce the final assemble difficulty, the axial bearings are utilized for preventing the rotor from touching the stator, and then the two rotor can be installed one by one. Figure 4-19 presents the diagram of the final assembly.

4.4 Prototype Test

The assembled prototype should be measured to investigate the performance and to validate the FEM results. In this section, the measuring results, the comparison with the calculation and the source of the errors are illustrated.

4.4.1 Resistance Measurement

Table 4-1 Measured phase resistance

Phase	A	B	C	Average	Calculated
Resistance Ω	0.134	0.139	0.137	0.137	0.129

The LCR meter is utilized for the resistance measurement with DC voltage source. The measured phase resistance at room temperature 25 °C is listed in Table 4-1. Compared with the calculated value the source of error may come from the resistivity diversity, welding spot resistance, and the discrepancy about the length calculation of the phase winding.

4.4.2 Inductance Measurement

The inductance is tested by rotating the rotor with 1 mechanical degree while the applied source current is of the rated frequency (140 Hz). Figure 4-20 illustrates the measured self and mutual inductances of three phases. The measured performance basically verified the calculation results. Compared with the calculated waveform, the misalignments may come from the uneven airgap caused by the manufacturing and the leakage inductance of the end winding and the outgoing lines.

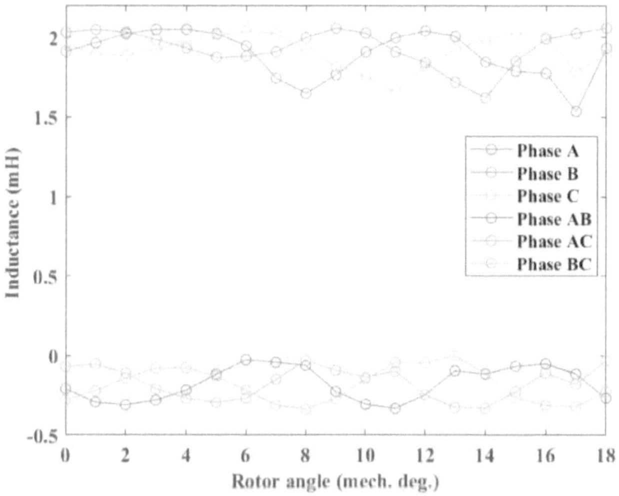

Figure 4-20 Measured self and mutual inductances.

4.4.3 Back EMF

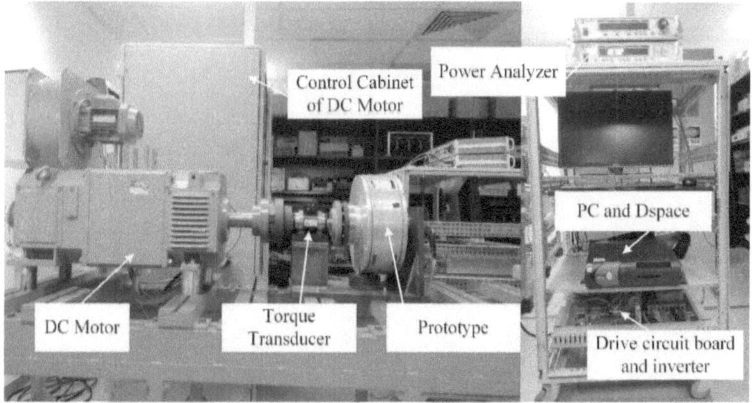

Figure 4-21 Test platform.

As shown in Figure 4-21, the measurement of the prototype has been carried out in a test rig developed by UTS CEMPE. The DC motor (DMP-4C from ABB) with the four-quadrant operation ability is utilized as the prime motor for the no-load voltage test and load test. The Magtrol TM 313 torque transducer is used for the output torque and speed measurement.

Driven by the DC motor, the no-load induced voltage of the prototype, i.e. back EMF is tested at rated speed 400 rpm with an oscilloscope. Figure present the test back EMF of each phase at 400rpm. The effective value of the measured no-load voltage waveform is 151.2V, compared with the computed 157.9V, the error is little which proves the accuracy of the finite element model.

However, due to the impaired strength of the adhesive which may be caused by the oxidation, the magnets and core components of the rotor fell out after the no-load test. After the reassembly, there is the degradation of the back EMF caused by the demagnetization in the repair process. As presented in Figure 4-23, the effective value of the phase voltage has degenerated to 142.6V. The decreased property of the magnets also leads to the reduction of the load performance in the following load test.

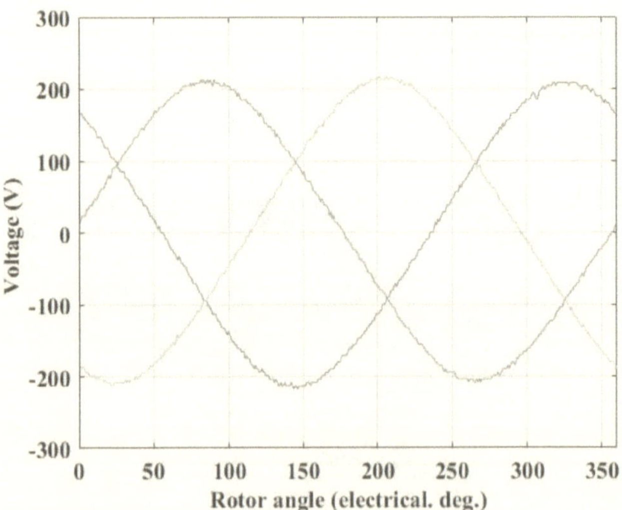

Figure 4-22 Measured no-load phase voltage waveform at 400rpm.

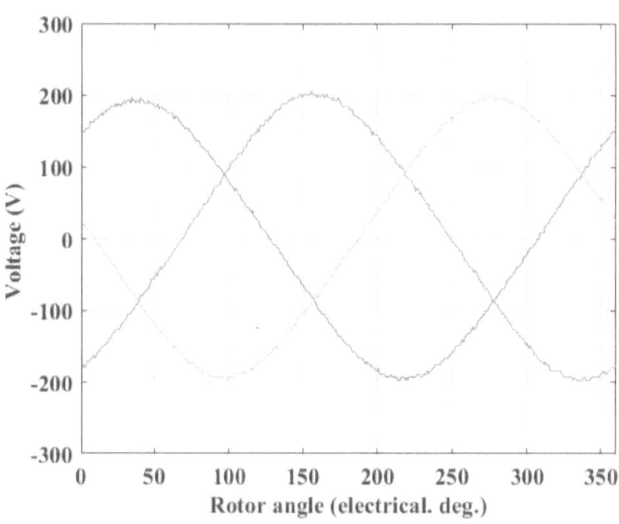

Figure 4-23 Measured no-load phase voltage waveform at 400rpm after the reassembly.

4.4.4 Loading test

As presented in Figure 4-21, The load test of the prototype is performed with a two-level inverter fed motor drive system. A dSPACE DS1104 PPC/DSP control board is utilized to implement the real-time algorithm coding using C language. The DC motor is operated in the electricity generation mode. The field-oriented control algorithm is applied with id =0. Due to the time limitation of the current power level of the drive circuit and power supply, and consideration of the mechanical stability, the load test of the motor is conducted with the low load condition at 400 rpm. The test result is presented in Figure 4-24. The load performance simulation is conducted after the correction considering the performance degradation, where there is still the misalignment. The main reason for the discrepancy of the simulated torque and tested values is the mechanical loss caused by the friction of the four bearings. The same reason causes the misalignment of the simulated and measured efficiency as illustrated in Figure 4-25.

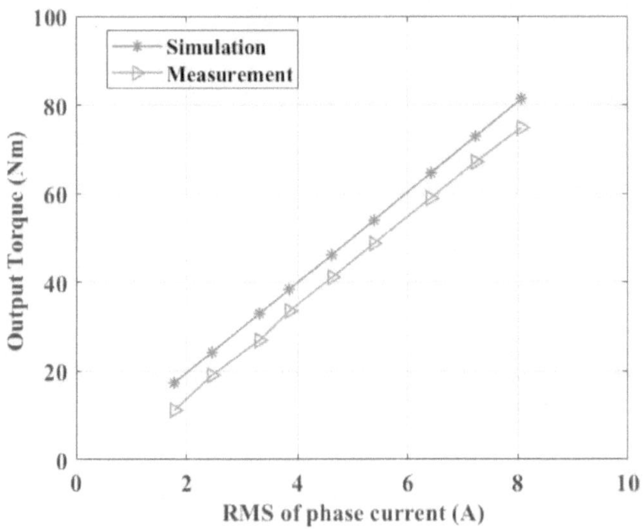

Figure 4-24 Output torque versus phase current.

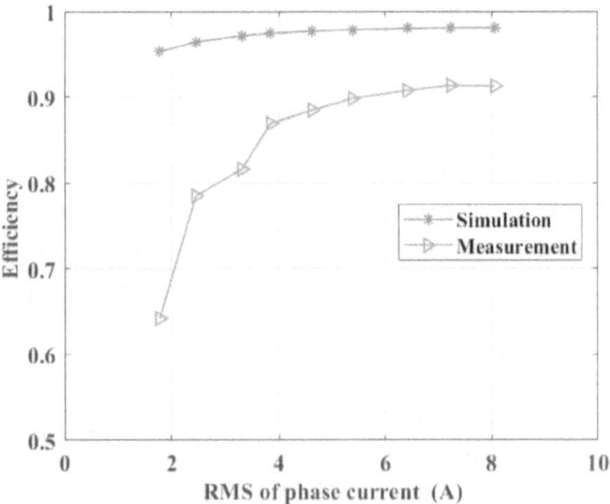

Figure 4-25 Efficiency comparison.

4.5 Optimization

According to the investigation of the electromagnetic design, prototyping and performance measurement on the dual rotor axial flux motor with grain-oriented silicon steel, the improvement scheme is proposed, which includes the mechanical design optimization and electromagnetic design optimization. The optimization of the mechanical design aims to increase the mechanical strength and stability of the design. The optimization of the electromagnetic design is conducted for the enhancement of the output performance.

4.5.1 Optimization of the Mechanical Design

To prevent the falling-off of the rotor core components, PMs and even the stator core components caused by the axial force and vibration, mechanical fixation can be applied for fastening the components on the carriers. The screw fixation can be adopted for the rotor core and carriers through the center hole punched on the silicon steel sheets. The concavity plate on the radial ends of the PMs can be designed for mechanical fixation on the rotor carrier with aluminum circular plates along the circumferential direction. Industrial adhesive can also be applied simultaneously for auxiliary stabilization.

To enhance the mechanical stability of the stator cores and winding components can be cast together with the epoxy resin which can also reduce the thermal resistance. For the fixation of the stator, the reference [1,2] offers a method with polyetheretherketone or polyoxymethylene plate from the axial ends of the stator instead of the fixing plate used in this prototype. The referred fixation method offers better mechanical stability and assembly precision which can be adopted for future prototyping.

As shown in the loading test results, the mechanical loss caused by the friction of the bearings made up a great proportion of the total loss. Even the application of the axial bearings relieves the difficulty of the assembly, the imbalanced axial force between the stator and rotors caused by the asymmetry air gap increases the frictional loss. Therefore, the mechanical installation stability and precision of the rotor and stator carriers are also important. Effective fixation of the carrier with high strength and precision can relieve the application of the axial bearings. Radial bearings with stop screws can be applied in the future for replacing the general radial bearings for preventing the axial movement of the rotor carrier.

4.5.2 Optimization of the Electromagnetic Design

From the aspect of the electromagnetic design, the current design has a problem with the large air gap which reduced the concentrated flux of the rotor design. Compared with the reluctance torque, the PM torque takes up the main part of the output torque. This is because the reluctance of the flux path of the current design in the direct-axis and quadrature axis is close. In other words, there is no obvious difference of direct axis and quadrature axis of the flux path. For increasing the ratio of the reluctance torque in the output torque, the V shape PM design and back silicon steel are required. This kind of improvement is not discussed in this research, while the following optimization is based on the current design and aims to the output performance while minimizing the mass of the motor design.

As analyzed above, the design parameters of the rotor design are selected first for the optimization for enhancing the PM torque output. Considering the mechanical design improvement, the optimization takes the design with the narrower air gap into account. Meanwhile, in view of the input voltage and current, the coil number and axial length of stator teech are also selected. The design parameters and variation bounds are presented in Table 4-2.

Table 4-2 Selected design parameters for the optimization

Par.	Symbol	Unit	Lower bound	Upper bound
Axial length of stator teeth	Hst	mm	30	110
Axial length of rotor teeth	Hrt	mm	12	18
Lower inner width of rotor teeth	Wrti	deg	4	7
Lower Outer width of rotor teeth	Wrto	deg	4	7
Upper inner width of rotor teeth	Wrti1	deg	4	7
Airgap length	Hgap	mm	1.1	2.1
Number of coils of each slot			10	40

The optimization model is expressed as

$$obj. \begin{cases} \max\ f_1 = T_{m1} \\ \min\ f_2 = M_{tot} \end{cases}$$

$$s.t. \begin{cases} g_1 = P_{cu} \leq P_{lim}, g_2 = H_{axial} \leq 130, \\ g_3 = U_m \leq \dfrac{U_{DC}}{\sqrt{3}}, g_4 = T_{m2} \geq \dfrac{T_{m1}}{4} \end{cases} \tag{4-1}$$

where T_{m1} and T_{m2} are the maximum output torques at speed of 400 rpm and 1600 rpm respectively, H_{axial} is the axial length of the design, P_{cu} copper loss considering the DC resistance, M_{tot} is the total mass U_m is the amplitude of the phase voltage, U_{DC} is the DC voltage of the inverter which is set as 600V. Since the copper loss dominates total loss of the motor in wide operation conditions, the cases with different limited copper loss P_{lim}, i.e. 1 kW,1.5 kW, and 2kW are discussed for the optimization.

Table 4-3 Optimal solutions of similar weight with the initial design

Copper loss limit	1kW	1.5kW	2kW
Mass (kg)	35	35	34.9
Maximum output torque at 400rpm	672	834	1006
Maximum output torque at 1600rpm	167	186	210

For the optimization conduction, the Kriging model is utilized for the performance estimation of the flux, inductance, and multi-objective GA is adopted as the optimization algorithm. The Pareto front of the optimal results is presented in Figure 4-26. We select the outer rotor radial flux motor in reference [3] of the chapter as a baseline design. The motor maximum output torque is 480 Nm at 400 rpm and the total active weight is 23.3 kg, while the copper loss is 1.7 kW. We select the optimal designs of 1.5 kW copper loss for comparison. Supposing the cooling condition is of the same with the baseline design, the designs of maximum torque larger than 480Nm while the mass lower than 23.3 kg can be selected as the designs with improved performance. The optimal designs with other copper limit should be considered with different cooling condition in the future works.

The maximum output torque at speed 400 rpm and 1600 rpm, and the weight of the solutions are illustrated in Figure 4-27. In the optimal solutions, the optimal solutions of similar weight with the initial design in Chapter 3 are listed in Table 4-3. The optimal solution of the 1.5 kW copper loss limit in the operation range shows 44 Nm higher output torque ability than the initial design. With a 2 kW copper limit, the same with the initial design, the maximum output torque at 400rpm increases to 1016 Nm.

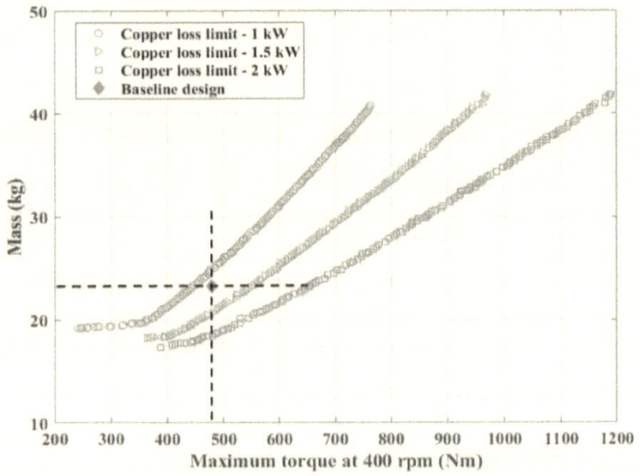

Figure 4-26 Optimization results with different copper loss limits.

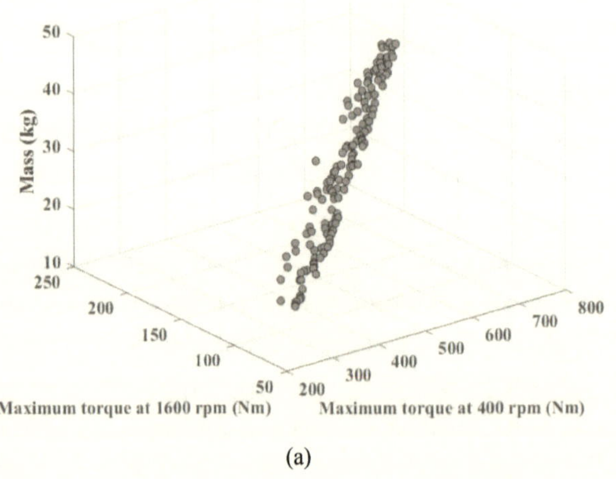

(a)

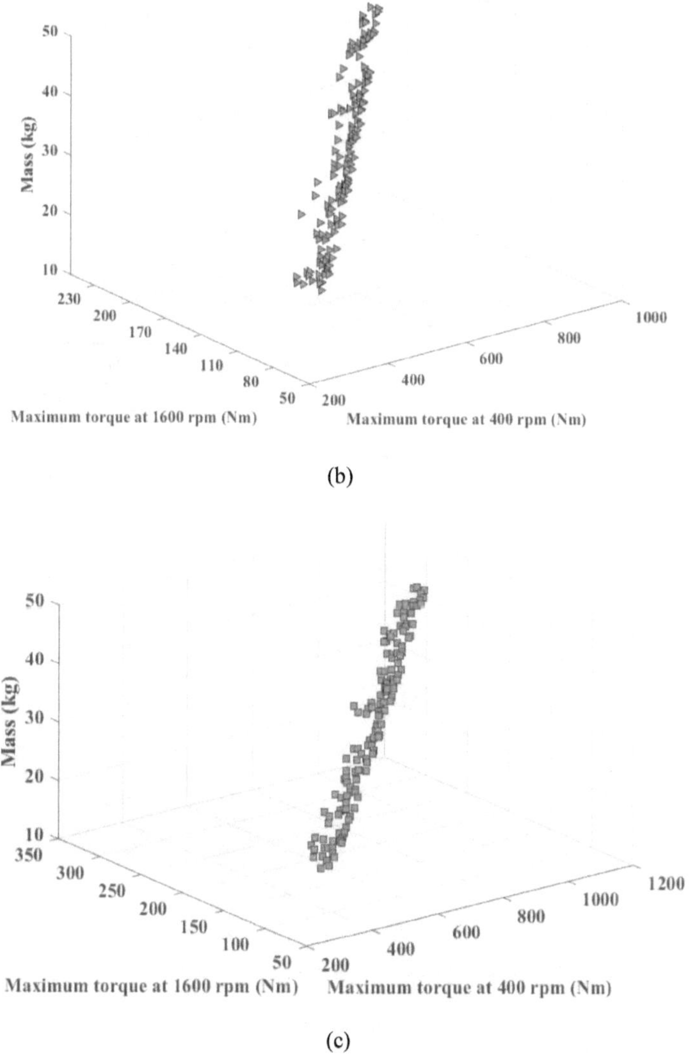

(b)

(c)

Figure 4-27 Mass, maximum torque at 400 rpm and 1600 rpm of the optimal designs with the maximum copper loss limit of (a) 1kW, (b) 1.5kW, (c) 2kW.

According to the design specifications, we select the optimal design with the copper loss limit of 2 kW and the maximum output of 500Nm for the performance verification and comparison with the design in reference [3]. The design parameters and the total mass are listed in Table 4-4 and the optimized rotor design is illustrated in Figure 4-28. Compared with the initial design shown in Figure 3-7, the optimal topology design parameter Wrti1 is enlarged obviously. In addition to the enhanced electromagnetic performance, the enlarged Wrti1 can also benefit mechanical design stability due to the increased contact area of the rotor core with carrier and reduce the manufacturing difficulty.

Table 4-4 Comparison between the selected optimal design and initial design

	Hst	N	Wrti	Hgap	Wrti1	Wrto	Hrt	Mass
Unit	mm		deg	mm	deg	deg	mm	kg
Initial design	80.8	28	0.8	2.1	6	5	13	35.1
Optimal design	45	20	6.2	1.17	4.7	6.4	12.7	19.1

Figure 4-28 Rotor component shape of the selected optimal design.

Figure 4-29 shows the efficiency map of the selected design, and Figure 4-30 illustrates the specific loss distributions. Compared with the design in reference [3] of 23.1kg weight, 480Nm maximum torque, and 1200 rpm maximum speed, the selected design shows higher speed expansion ability, less active mass, and comparable efficiency in the operation region, which proves the potential of the dual rotor axial flux motor for the wheel motor design.

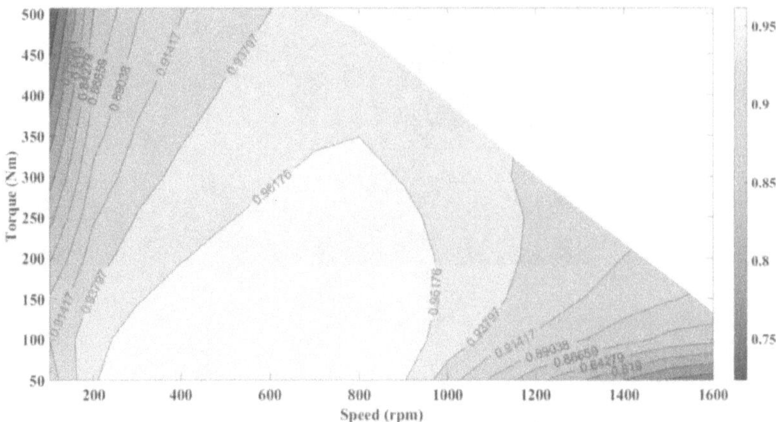

Figure 4-29 Efficiency map of the optimal design.

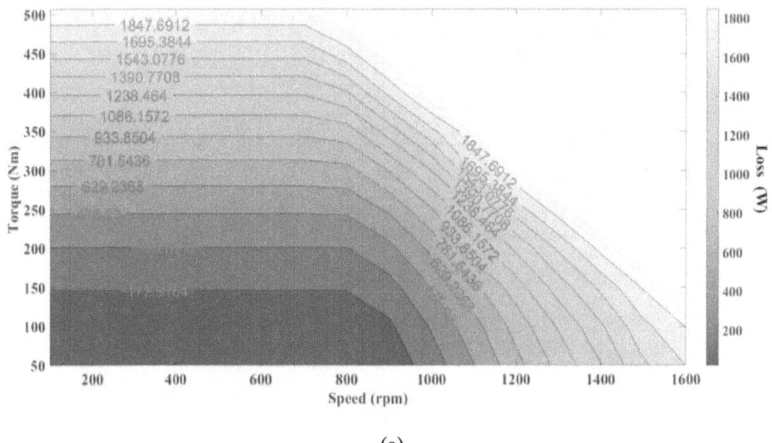

(a)

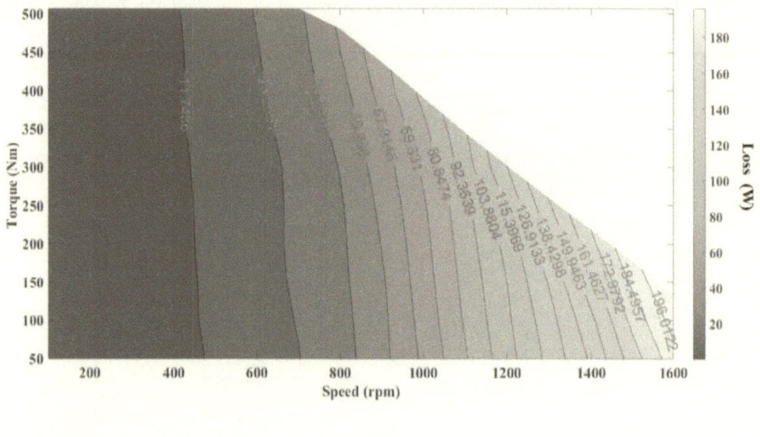

(b)

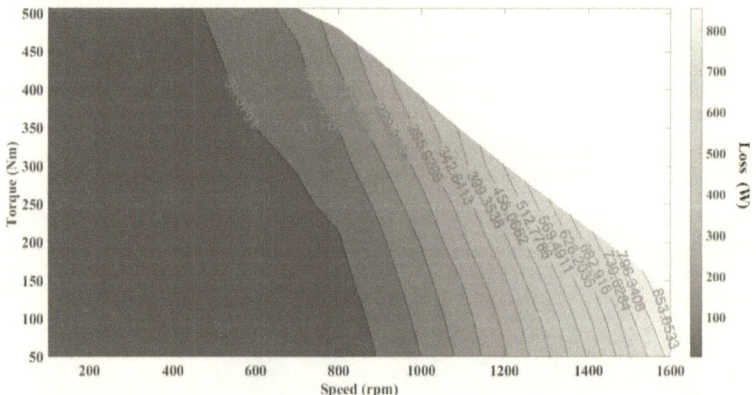

(c)

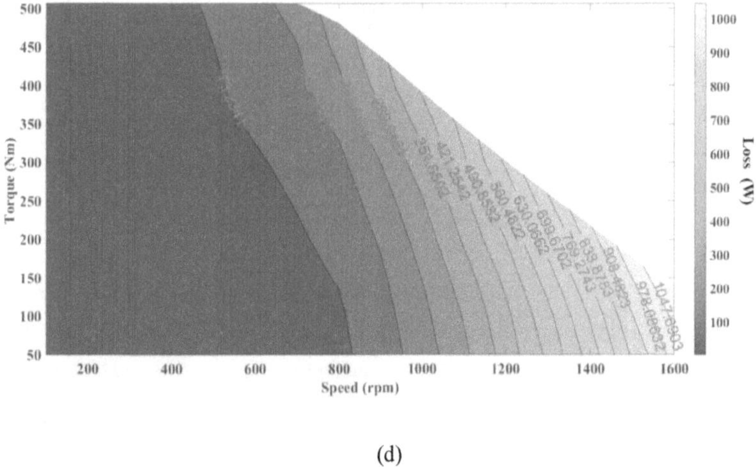

(d)

Figure 4-30 Map of the loss (W) (a) copper loss, (b) core loss, (c) PM loss, (d) PM loss and core loss.

4.6 Conclusions

The manufacturing techniques for the designed dual rotor axial flux in-wheel motor is presented in the chapter. The prototype test results basically verified the computed performance. With the accumulated manufacturing experience in the prototyping, the mechanical design improvement scheme is presented for enhancing the mechanical stability of the future prototyping. Based on the verified model, the multi-objective optimization of the output torque at rated speed and mass is conducted considering with considering the performance in the high-speed flux weakening region, limitation of the design space and various copper loss. The output ability of the optimal design is enhanced of the same weight with the initial design. For performance comparison with an in-wheel motor of the outer rotor in the referenced paper, one of the optimal designs in the Pareto front selected for performance verification with the finite element model. The comparison shows the advantage of the optimal design in speed expansion and weight reduction while possessing comparable efficiency in the operation region.

Chapter 5 Topology Optimization for Electrical Machines

5.1 Introduction

As introduced in Chapter 2, due to the advances of the ideology, topology optimization techniques have been initiated and researched extensively in the mechanical field and also extended to other engineering fields such as the thermotic, acoustics, electromagnetics, etc[1]. Various methods have been proposed which mainly include homogenization method, density method, level set method, discrete method, and level set method, etc.

Compared with the development of topology optimization in the field of mechanical design, research in the field of electromagnetic engineering is relatively scattered. Furthermore, since the design optimization of electrical machines is only a branch of the electromagnetic engineering field, the relevant topology optimization research for electrical machines is scarcer. This introduction section of the chapter reviews the general points of the specific optimization methods. In addition, their application in the design optimization of electrical machines and the relevant design problems such as the electromagnetic actuators is mainly investigated. Finally, the summary and research details in this chapter are provided.

5.1.1 Literature Review on the Topology Optimization Methods for Electrical Machines

5.1.1.1 Homogenization Method (HM)

Topology optimization technology is regarded as originating from the development of the homogenization method [2]. By applying the homogenization method, the design variables or elements are regarded as formed with composite structures. Some microstructures from the references [2-5] are shown in Figure 5-1 to Figure 5-4. First of all, the application of the microstructure is considered to be with a physical meaning. The material property can be changed by varying the sizing parameters of the composite structures. The layout of the design domain is then transferred into the problem of the material property optimization of elements with the composite structures. Each composite structure can be defined by several parameters for representing its shape and size of the empty and solid parts. The material property of each microstructure is expressed by interpretation. After the optimization, the optimal topology of the design domain will be generated where the void parts are composed of the microstructures with large cavities while the solid is filled with the microstructures with little holes.

The research of the HM in electromagnetic design problems can be found in [3], in which the calculation of electromagnetic field parameters and their sensitivities are conducted. Later works from the same research group include the application of the method for vibration reduction in the reluctance machine and actuator [4, 5], and improving the force density in the actuator [6, 7].

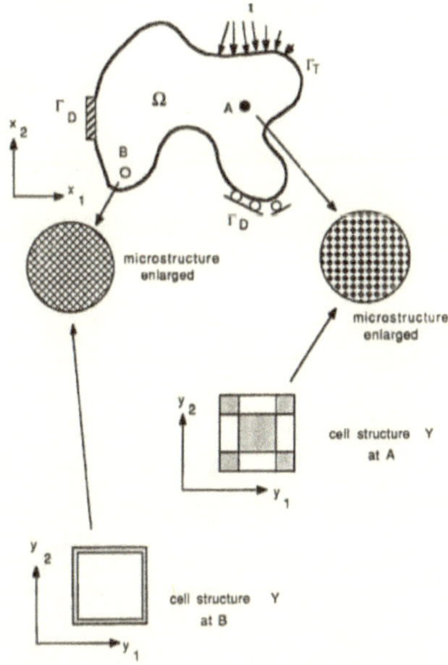

Figure 5-1 Microstructure in [2].

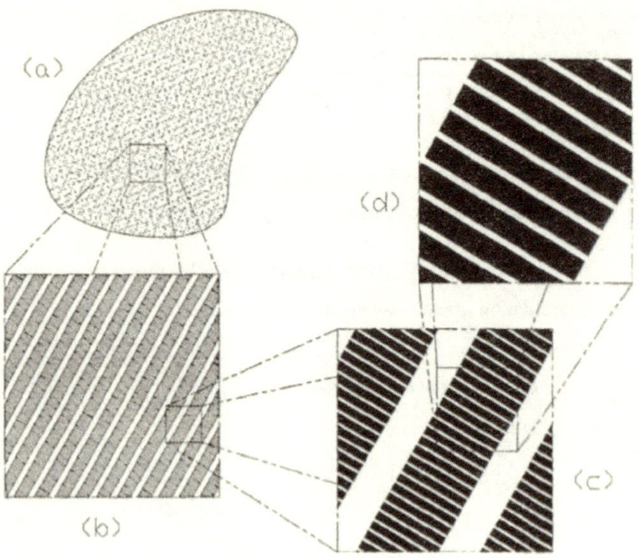

Figure 5-2 Microstructure in reference [3].

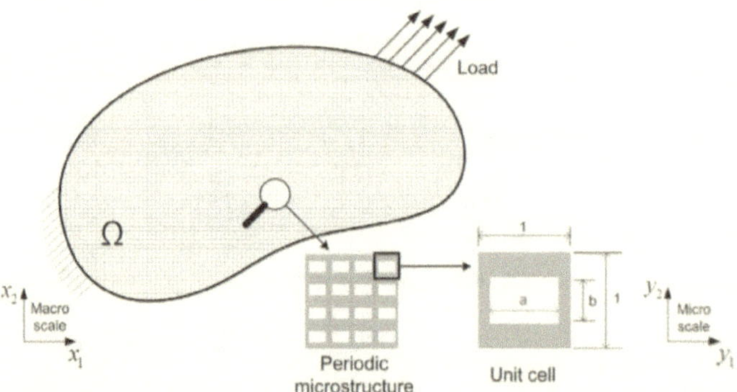

Figure 5-3 Microstructure in [4].

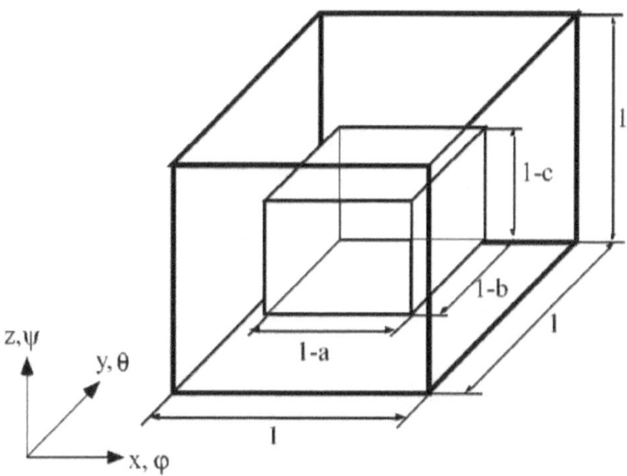

Figure 5-4 Microstructure in [5].

First of all, the HM transferred the discrete optimization problem into a continuous problem. The derived performance and sensitivity calculation made it possible for the application of gradient-based algorithms for optimization, which decreases the computation burden. The application of the composite structure and the interpolation method is recognized with the physical meaning. However, the application of the HM is not widely applied in the design optimization problems for the electrical machines. Some general drawbacks for the application of this method include the application of the deterministic algorithm for the high dimensional continuous problem solving which usually leads to the local minimum. Particularly, the design dimension is extremely high because of the design parameters of the microstructure and element number. In the final optimal solution, not all the composite structure will converge to the void or solid which means, which means the microstructure with both void and solid materials will exist. Even though with physical meanings, the optimal design with an intermediate structure can hardly be manufactured. Due to the manufacturing difficulty, post-processing techniques may be required for eliminating these elements.

5.1.1.2 Density Method

The density method also is known as simplified isotropic material with penalization (SIMP) approach tries to simplify the application of the microstructures by utilizing the density powered by an exponent penalization factor for interpolation of the material property [9, 10]. As introduced in [11], various interpolation methods have been developed. For example, the widely applied one in structural design which can be expressed as

$$E(\rho_i) = \rho_i^p E_0 \tag{5-1}$$

where E_0 is the Youngs Modulus of the solid, ρ_i is the element density i.e. design parameter and p is the penalization factor. As shown in Figure 5-5, the variation space [0,1] of the design parameter realizes the material property continuously changing from vacuum to solid material.

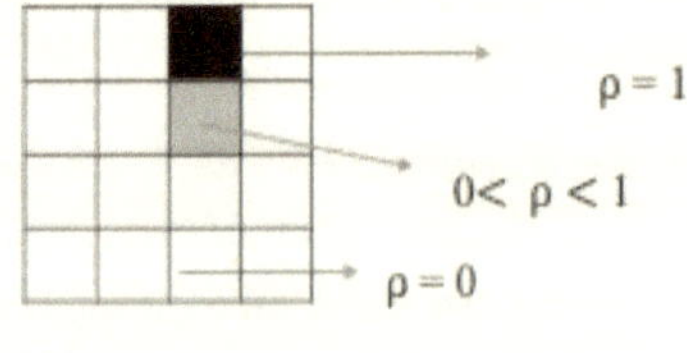

Figure 5-5 Elements with different density.

This method was first imported into the electromagnetic design problems by David Lowther [12] with the interpolation of□ $\mu r_e = \mu r_{air} \mu r_{fe}^{\rho_e}$ for the component of ferromagnetic material optimization, where μr_{air}, μr_{fe} are the relative permeabilities of air and ferromagnetic material ρ_e is the element density, and μr_e is the relative permeability of the element. Similarly, the material property in each element is interpolated by the equation which realized the continuous transition from air to iron. Compared with the utilization of the composite structures in the HM method, the design parameter definition is simplified with an effectively reduced dimension. The design dimension equals to the element number if FEM is applied. This simple and direct processing method also predigests the performance and sensitivity computation, which attracts the interest of researchers rapidly and obtain various applications for electromagnetic design problems [13]. However, the investigation of the density method in the design optimization of electrical machines is rare. A specific application can be found in [14] for the automatic design of the rotor for the reluctance motor by the derivation of the mutual energy. A detailed torque profile optimization for the reluctance motor by the density method is presented in [15]. In this research, the sensitivity of the objective and constraint including average torque and torque ripples with respect to the ferromagnetic materials of the stators and rotors are derived. The optimization is conducted under the linear condition of the ferromagnetic component and the results show the effectiveness of the method. The optimization of the output torque of an interior PMSM (IPMSM) by the density method is researched in [16]. The performances of average torque, torque ripple and cogging torque included in the optimization model are improved.

Even the merits of density method are obvious in contrast to HM, there are some general problems of the method, specific problems in the design optimization of electrical machines, and some developed solutions should be mentioned here. Compared with the microstructure application in HM, the fictitious material density is lack of physical meaning. As the aforementioned problem in the HM, the composition structures can hardly converge to the vacuum or solid components in the final solution. Similarly, the element with middle density (not 0 or 1) can be inevitable. This is also the point that the density method is debatable. On one hand, compared with HM, the density method offers a more direct way of modeling the topology problem. However, due to the unavoidable element with intermediate density, the optimal solution with a physical significance can hardly be obtained. Nevertheless, there is also the argument that the intermediate density element can be replaced with composite structures. An interesting work can be found in [8] for verification of the thoughts, which takes advantage of density method for the topology optimization of a linear machine, and replaces the elements of middle density with the microstructure for obtaining the optimal design with a physical meaning. The optimal solution is shown in Figure 5-6.

Figure 5-6 Optimization result in reference[8].

It has been pointed out that due to the application of the FEM, there is a mesh dependency problem in which the optimal solution has a high dependence on the meshing density. There is also the possibility of the checkboard design. These problems have been widely investigated in the mechanical field. Techniques like filters or geometrical restrictions have been verified. The specific techniques have been categorized into three kinds, i.e. 1, 2, 3 field approaches in [1]. However, since these methods are investigated in the background of mechanical design problems, the universality in design problems of other fields is unknown. Due to the transformation from a discrete problem into a continuous one and optimization by the gradient-based algorithms, there is also the general problem of the local minima.

As mentioned before, the design optimization of electrical machines is a multi-material, multi-objective problem. The FEM for electrical machines is also a nonlinear problem considering the saturation of ferromagnetic material. As mentioned, in the electromagnetic design of electrical machines, there are generally 3 to 4 materials included which are air, Ferromagnetic material, conductor material and PM. Currently, most investigation of the density method is about optimizing the layout of ferromagnetic material and air. Considering more materials at a time means more complicated interpolation for different properties which aggravates the solving difficulty. Some pioneer works can be found in [17, 18] for the multi-material optimization of actuators.

Considering the complexity, stability and computation cost of the optimization method, the multi-material optimization techniques still require systematic and detailed research.

In addition, the electrical machine is an energy conversion system, which implements the transition between electricity and mechanical power by the magnetic field. As aforementioned, various performances should be considered in multi-objective optimization. For the application of the density method, the effective sensitivity analysis method should be emphasized. Based on the existing research, the sensitivity analysis methods can be classified into two categories, i.e. direct method [15] and adjoint variable method [19]. The direct method is developed from the parameters and performances of the field with more consideration of the physical significance, while the adjoint variable method offers a straightforward and mathematical construct for the sensitivity derivation. However, the effectiveness of both of the methods still requires intensive research on the optimization of electrical machines. The research on the direct method also composes one of the important parts of this research which is introduced in the following sections of the chapter.

5.1.1.3 Level Set Method

The level set method is initially developed in [20] for calculation of the evolution of the interfaces. The early works of the level set method based topology optimization can be found in [21, 22]. By introducing in level set function (5-2), the material in the design domain can be divided according to the function values. The boundary is obtained by the interface of the zero-level set and the design domain.

$$\phi(\mathbf{x}) \begin{cases} = 0 \text{ if } \mathbf{x} \in \partial\hat{\Omega} \\ > 0 \text{ if } \mathbf{x} \in \hat{\Omega} \\ < 0 \text{ if } \mathbf{x} \in (\Omega \setminus \hat{\Omega}) \end{cases} \tag{5-2}$$

The evolving of the level set function is by solving the Hamilton-Jacobi equation

$$\frac{\partial\phi}{\partial t} + V \, |\nabla\phi| = 0 \tag{5-3}$$

where time t is of no physical meaning but the iteration time of the optimization, V is the normal velocity. Defining the velocity by the shape sensitivity with respect to the objectives, the evolution of the level set function is driven for the optimum. Since the optimization is driven by the movement of the level set function, the adjoint variable method is normally applied for calculation of the normal velocity.

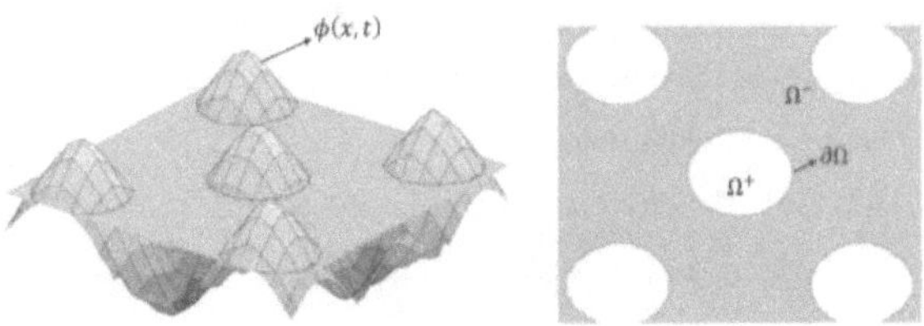

Figure 5-7 Level set function and domain separation on zero level set plane.

Although the research and application of the level set method in the design optimization of the electrical machine are relatively late, we can see a fast increase in the last ten years which makes it with the most referenced works in the field. Earlier works can be found for the rotor topology optimization of reluctance motor where there is only one type material i.e. ferromagnetic material, while the torque profile is of the high sensitivity of the rotor design [23-25]. Others include maximizing the magnetic field energy in the actuator and reducing the torque ripple of the PMSM by optimizing the ferromagnetic materials. The application of the level set method for two materials then spread to the multi-material problems with a multiphase level set method [26, 27]. In the research of [27], the relative permeability and reluctivity are interpolated for evolving of the material distribution on the rotor of a PMSM. The phase-field method has been taken advantage of controlling the shape of the optimal solution. A series of work on the multi-material optimization with the multiphase level set method is published by Putek for the shape optimization of a surface-mounted PMSM (SPMSM) [28-33]. From one hand, we can find the advances of the researches which try to cover as many performances as possible in the optimization models. In particular, to the knowledge of the author, the combination with uncertainties analysis method (PCE) and multiphase level set method for robust topology optimization, which is currently a hot topic in the structural optimization fields [1], is the only research in the topology optimization for electrical machines [32]. On the other hand, even though we can find the improvement of objective performances, the shape variation of the components in the optimization domain is not obvious. This may be because of the complexity of the problem and the restriction of the current methods.

From the aspects of the method, the application of the level set function for solving the shape variation in the optimization allows the splitting, merging and hole degeneration of the initial design. As one of the merits, the optimal solution achieved by the level set method normally has smooth edges between the component of different materials, as presented in Figure 5-7.

There are also some drawbacks to the method that should be noted. Because of the application of gradient-based optimization algorithms, there is the general problem of convergence to the local optimum. Due to the lack of nucleation mechanism, the generation of holes by the level set method requires the topological derivative or forcing generation method. Generally, the optimal results highly rely on the initial settings. There is also the difficulty in solving the Hamilton-Jacobi equation for updating the level set function. Similar material interpolation with density method is used for interfaces for sensitivity calculation. Therefore, the problems in the density method application such as the optimal design with intermediate density also challenge the application of the level set method.

5.1.1.4 Discrete Method

The discrete method, or more widely known as the ON-OFF method in the design optimization of the electromagnetic design problem uses binary 0 and 1 to represent the solid material existing or not. Compared with the optimization methods by transferring the problem into a continuous one, this method utilizes the discrete nature of the problem. In contrast to the application of the deterministic gradient-based optimization methods for fast extreme solution searching, this kind of method takes advantage of the stochastic intelligent optimization methods such as evolutionary algorithms or genetic algorithms with abilities for the global optimum. Also, there are not the problems of the solution with a fictitious optimal solution which occurs by applying the continuous method such as the gray design with intermediate density by the density method.

It has been pointed out that there will be the possibility of the non-existence of solutions and mesh dependency if the ON-OFF method is applied. In the optimization of electrical machines, if the elements in the FEM are directly optimized as the design variables, the dimension of the design variables can be extremely large. Due to the application of the intelligent algorithms, the iteration steps for the global optimum searching is also more than the deterministic algorithms. Both design dimensions and iterations will cause a huge demand for the FEM calculation. Therefore, the computation burden of the method can be very heavy. The numerical instability problem such as the checkboard patterns can also easily happen by utilizing the discrete method.

Current efforts about the application of the ON-OFF method are introduced below. We can see the efforts on taking advantage of the computation power of cluster computation or parallel computing [19]. It is noted that the design takes the PM as into consideration with air and ferromagnetic material. Meanwhile, to reduce the design dimension, the coarse design grid can be firstly utilized for the first-round optimization. Then the solution is divided into more detailed elements for further optimizations [34, 35]. For the reduction of computation cost by reducing the iteration numbers, we can see the work with effective sensitivity calculation for assisting the convergence of the algorithms [35, 36]. In particular, the ON-OFF sensitivity method was applied for the torque profile optimization of IPMSM in [37, 38], in which the sensitivity is calculated with respect to the reluctivity with the nodal force method and adjoint variable method. The approximation model is also used in the [39] for reducing both of the influences of computation burden and checkboard design. For overcoming the checkboard design, the efforts can also be seen in the design and application of the filters such as the geometry constraints [36, 40]. Other works on the PM optimization can be found in [41].

5.1.2 Work in This Chapter

As reviewed above, despite the various methods that have been applied in the topology optimization of electrical machines, the research on each of the methods is limited. Although the topology optimization has shown higher freedom compared with parametric design, the automaticity for the design optimization of electrical machines is still low compared with the conceptual topology bring-up in the structural design. The current topology optimization methods still required a well-designed initial structure. Therefore, from the view of the author, the works should be categorized into the shape optimization of electrical machines with the topology optimization method. The research on the topology methods for electrical machines should be conducted extensively for a general approach to the optimization problem while increasing design optimization freedom and automaticity.

This research presents an effort to develop the density method for optimizing ferromagnetic components in electrical machines. As presented in the limited number of reference papers applying the density method for electrical machine optimization, further research is still required on the optimization framework establishment based on a general method for electromagnetic performance derivation and its sensitivity analysis. The rest of this chapter is organized as follows. Section 5.2 describes the electromagnetic problem and solution process of FEM. Considering the sensitivity analysis, Section 5.3 derives the parameter and performance calculation methods in the optimization problems of electrical machines. The parameters and performance include the energy stored in the magnetic field, flux linkage and torque. The calculation of energy stored in the magnetic field is fundamental for the other calculations. Section 5.4 presents the topology optimization based on the density method, including the interpolation method, parameter and performance sensitivity analysis, algorithms, optimization framework, and post-processing method. To validate the effectiveness of the proposed optimization method, two design examples are reported and discussed in Section 5.5, followed by the conclusion drawn in Section 5.6.

5.2 Fundamentals of Finite Element Model for Electrical Machine Analysis

For electrical machine design, the magnetic field is analyzed for given sources and boundaries by using the magnetic vector potential, $\mathbf{A}$, defined by

$$\mathbf{B} = \nabla \times \mathbf{A} \tag{5-4}$$

where $\mathbf{B}$ is the magnetic flux density. According to the Maxwell equation, the magnetic field strength, $\mathbf{H}$, is related to the current density, $\mathbf{J}$, by

$$\nabla \times \mathbf{H} = \mathbf{J} \tag{5-5}$$

The relationship between the flux density, $\mathbf{B}$, and the magnetic field strength, $\mathbf{H}$, is determined by the material characteristics. In engineering practice, we assume

$$\mathbf{H} = \frac{\mathbf{B}}{\mu} = \nu \mathbf{B} \tag{5-6}$$

where $\mu = \mu r \mu_0$ is the magnetic permeability, μr the relative permeability of the material, and $\mu_0 = 4\pi \times 10^{-7}$ the permeability of the vacuum. In a ferromagnetic material, μr is a nonlinear parameter determined by the normal magnetization curve if the magnetic hysteresis is ignored. where magnetic reluctivity ν is the inverse of permeability μ.

Combining (5-4)-(5-6) and applying the Coulomb gauge

$$\nabla \cdot \mathbf{A} = 0 \tag{5-7}$$

one obtains

$$\nabla \times (\nabla \times \mathbf{A}) = \nabla (\nabla \cdot \mathbf{A}) - \nabla^2 \mathbf{A} = -\nabla^2 \mathbf{A} \tag{5-8}$$

$$\nabla^2 \mathbf{A} = -\mu \mathbf{J} \tag{5-9}$$

For a two-dimensional field in the Cartesian coordinate system, where the magnetic vector potential and current density have the components only in the z axis, (5-9) can be simplified as

$$\nabla^2 A_z = \frac{\partial^2 A_z}{\partial x^2} + \frac{\partial^2 A_z}{\partial y^2} = -\mu J_z \tag{5-10}$$

To achieve the unique solution to a problem, the specified condition should be known at each point on the boundary. When the potential values at the points on the boundary are specific, we call this a Dirichlet condition (or first boundary condition). Particularly, if the value of potential is zero, it is named as homogeneous Dirichlet condition. When the normal derivative of the potential is specified, we specify a Neumann condition (or second boundary). Similarly, when the normal derivative is zero, it is called homogeneous Neumann condition, or natural boundary condition. The periodical boundary condition is also artificially set according to the periodic distribution of the magnetic field in electrical machines for decreasing the computation burden.

Combined with the partial derivative equation and boundary conditions, the boundary-value problem can be summarized as

$$\begin{cases} \dfrac{\partial^2 A_z}{\partial x^2} + \dfrac{\partial^2 A_z}{\partial y^2} = -\mu J_z \\[2mm] s_1 \quad A_z = A_{z0} \\[2mm] s_2 \quad \dfrac{1}{\mu}\dfrac{\partial A_z}{\partial n} = -H_t \end{cases} \tag{5-11}$$

where s_1, s_2 are the Dirichlet condition and Neumann condition, respectively. A_{z0} and H_t are the known values of the boundary conditions.

The FEM is a numerical procedure for achieving the solutions to the boundary value problems. The boundary-value problem is equated to a conditional problem of variation. A continuous domain is replaced by a number of subdomains in which the unknown function is replaced by interpolation functions and the energy and potential can be expressed. After the assembling of all the elements, the algebraic equations are obtained. Finally, the solution of the problem is achieved by solving the equations, and then the magnetic field quantities can be obtained successively.

Taking the 2-dimensional steady-state magnetic problem as an example, the conditional problem of variation, i.e. the extreme value problem of the energy functional can be expressed as

$$\begin{cases} \min W(A) = \iint_{\Omega}\left[\left(\dfrac{1}{2\mu}\dfrac{\partial^2 A_z}{\partial x^2} + \dfrac{1}{\mu}\dfrac{\partial^2 A_z}{\partial y^2}\right) - JA\right]\mathrm{d}x\,\mathrm{d}y + \int_{s_2} H_t A_z \mathrm{d}s \\[2mm] s_1 \quad A_z = A_{z0} \end{cases} \tag{5-12}$$

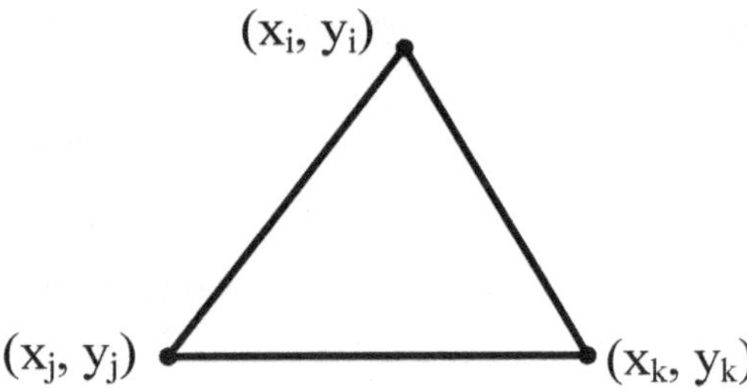

Figure 5-8 Triangular element.

As a widely used element, the triangular element is utilized for explaining the basic steps of the FEM. For a single triangular element as shown in Figure 5-8 in the cartesian coordinate system in which the vertices are numbered counter-clockwise, the vector potentials at these points are unknown variables to be evaluated. Then the vector potential inside of the element can be calculated approximately according to the linear interpolation function as

$$A_z = \frac{a_i + b_i x + c_i y}{2\Delta} A_i + \frac{a_j + b_j x + c_j y}{2\Delta} A_j + \frac{a_k + b_k x + c_k y}{2\Delta} A_k \qquad (5\text{-}13)$$

where the A_i, A_j, A_k are the vector potentials at the vertices and the coefficients are also known as the shape functions. The coefficients in the shape function can be calculated as below

$$\begin{aligned} a_i &= x_j y_k - x_k y_j \\ b_i &= y_j - y_k \\ c_i &= x_k - x_j \end{aligned} \qquad (5\text{-}14)$$

$$\begin{aligned} a_j &= x_k y_i - x_i y_k \\ b_j &= y_k - y_i \\ c_j &= x_i - x_k \end{aligned} \qquad (5\text{-}15)$$

$$\begin{aligned} a_k &= x_i y_j - x_j y_i \\ b_k &= y_i - y_j \\ c_k &= x_j - x_i \end{aligned} \qquad (5\text{-}16)$$

The symbol Δ is the area of the triangular element which can be expressed as.

$$\Delta = \frac{1}{2}\begin{vmatrix} 1 & x_i & y_i \\ 1 & x_j & y_j \\ 1 & x_k & y_k \end{vmatrix} = \frac{1}{2}\left(b_i c_j - b_j c_i\right) \tag{5-17}$$

The energy in the element is calculated according to the energy function. If there is no edge of the element on the second boundary condition or one edge on the natural boundary condition, the linear integral part in the equation can be neglected. Then the energy functional can be expressed as

$$W(A) = \iint_\Omega \left[\left(\frac{1}{2\mu}\frac{\partial^2 A}{\partial x^2} + \frac{1}{\mu}\frac{\partial^2 A}{\partial y^2}\right) - JA\right] dx\,dy \tag{5-18}$$

The energy of the element can be calculated as

$$W_{\text{mag,e}}\left(A_i, A_j, A_k\right) = \frac{1}{8\mu\Delta}\left[\left(b_i A_i + b_j A_j + b_k A_k\right)^2 + \left(c_i A_i + c_j A_j + c_k A_k\right)^2\right]$$
$$-\frac{J\Delta}{3}\left(A_i + A_j + A_k\right) \tag{5-19}$$

Taking derivatives with respect to the vector potential at the vertices

$$\begin{bmatrix} \dfrac{\partial W_{\text{mag,e}}}{\partial A_i} \\[2ex] \dfrac{\partial W_{\text{mag,e}}}{\partial A_j} \\[2ex] \dfrac{\partial W_{\text{mag,e}}}{\partial A_k} \end{bmatrix} = \frac{1}{4\mu_e\Delta}\begin{bmatrix} b_i^2 + c_i^2 & b_i b_j + c_i c_j & b_i b_k + c_i c_k \\ b_i b_j + c_i c_j & b_j^2 + c_j^2 & b_j b_k + c_j c_k \\ b_i b_k + c_i c_k & b_j b_k + c_j c_k & b_k^2 + c_k^2 \end{bmatrix}\begin{bmatrix} A_i \\ A_j \\ A_k \end{bmatrix} - \frac{\Delta J_0}{3}\begin{bmatrix} 1 \\ 1 \\ 1 \end{bmatrix} \tag{5-20}$$

If one of the edges of the element located at the second boundary condition, the derivative of the linear integral part of the equation can be expressed as

$$
\begin{bmatrix} \dfrac{\partial W^{s_2}_{mag,e}}{\partial A_i} \\[4mm] \dfrac{\partial W^{s_2}_{mag,e}}{\partial A_j} \\[4mm] \dfrac{\partial W^{s_2}_{mag,e}}{\partial A_k} \end{bmatrix} = \begin{bmatrix} 0 \\[4mm] \dfrac{H_t l_{jk}}{2} \\[4mm] \dfrac{H_t l_{jk}}{2} \end{bmatrix}
\tag{5-21}
$$

Similarly, the derivatives can be expressed as

$$
\begin{bmatrix} \dfrac{\partial W_{mag,e}}{\partial A_i} \\[4mm] \dfrac{\partial W_{mag,e}}{\partial A_j} \\[4mm] \dfrac{\partial W_{mag,e}}{\partial A_k} \end{bmatrix} = \frac{1}{4\mu_e \Delta} \begin{bmatrix} b_i^2 + c_i^2 & b_i b_j + c_i c_j & b_i b_k + c_i c_k \\[2mm] b_i b_j + c_i c_j & b_j^2 + c_j^2 & b_j b_k + c_j c_k \\[2mm] b_i b_k + c_i c_k & b_j b_k + c_j c_k & b_k^2 + c_k^2 \end{bmatrix} \begin{bmatrix} A_i \\[2mm] A_j \\[2mm] A_k \end{bmatrix} - \begin{bmatrix} \dfrac{\Delta J_0}{3} \\[3mm] \dfrac{\Delta J_0}{3} - \dfrac{H_t l_{jk}}{2} \\[3mm] \dfrac{\Delta J_0}{3} - \dfrac{H_t l_{jk}}{2} \end{bmatrix}
\tag{5-22}
$$

$$
\begin{bmatrix} \dfrac{\partial W_{mag,e}}{\partial A_i} \\[4mm] \dfrac{\partial W_{mag,e}}{\partial A_j} \\[4mm] \dfrac{\partial W_{mag,e}}{\partial A_k} \end{bmatrix} = \frac{1}{\mu_e} \mathbf{k}_e \mathbf{a}_e - \mathbf{f}_e
\tag{5-23}
$$

which has the same formation with the expression of the no edge on the second boundary condition or one edge on the natural boundary condition.

Once the element matrices are found for each element, they are used to form the global or system matrix in a process known as the assembly. Each element matrix has rows and columns corresponding to the nodes in the element. Suppose there are n nodes in total, build an $n \times n$ zero matrix $\mathbf{K}$ and an $n \times 1$ zero matrix $\mathbf{F}$. Then go through all the elements and add the $\mathbf{k}_e$ and $\mathbf{f}_e$ with the same column and row number to form the global matrix $\mathbf{K}$ and $\mathbf{F}$.

$$\begin{bmatrix} \dfrac{\partial W_{mag}}{\partial A_1} \\[2mm] \dfrac{\partial W_{mag}}{\partial A_2} \\[2mm] \vdots \\[2mm] \dfrac{\partial W_{mag}}{\partial A_n} \end{bmatrix} = \mathbf{KA} - \mathbf{F} = 0 \qquad\qquad (5\text{-}24)$$

where $\mathbf{K}$ is known as the stiffness matrix, $\mathbf{A}$ is the matrix of the vector potential, and $\mathbf{F}$ is the vector of excitations. According to the theory of calculating the extreme value for the function of many variables, the partial derivatives of W_{mag} with respect to vector potential are zeros. Then we can achieve

$$\mathbf{KA} = \mathbf{F} \qquad\qquad (5\text{-}25)$$

Until now, the basic process of the FEM for the steady-state magnetic field is presented. Moreover, to achieve the final solution, other necessary numerical process steps are still required including but not limited to the methods for the boundary condition, nonlinear solution, and magnetization models of the PM, etc. Due to the limitation of the space and the aim of introducing the candidate's work, the detailed description of these steps is not elaborated in this part. A more detailed introduction can be found in the references [42].

5.3 Performance Calculation

After the solution of the finite element analysis is finished, the vector potential at each node of the elements is achieved. For the analysis of electrical machines and design optimization, the performances of the electrical machines should be calculated by postprocessing of the known information of the finite element analysis. These performances usually include flux linkage, back electromotive force, loss, inductance, torque, etc.. The calculation of these characteristics according to the finite element analysis results is introduced in this section. Particularly, the calculation methods based on the energy information of elements are emphasized because of the sensitivity analysis requirement in the following sections for the topology optimization.

After the solution of the finite element analysis, the vector potential at each point is achieved, the flux density at each element can be calculated. The x and y components of the magnetic flux density can be calculated as

$$B_x = \frac{\partial A_z}{\partial y} = \frac{A_i b_i + A_j b_j + A_k b_k}{2\Delta} \tag{5-26}$$

$$B_y = -\frac{\partial A_z}{\partial x} = -\frac{A_i c_i + A_j c_j + A_k c_k}{2\Delta} \tag{5-27}$$

The magnetic energy of the element can be expressed as

$$W_{mag,e} = l_a \iint \frac{B_e^2}{2\mu_e} dxdy = \frac{1}{2\mu_e} l_a B_e^2 \Delta_e \tag{5-28}$$

where B_e, μ_e, l_a and Δ_e are the magnetic flux density, magnetic permeability, axial length and area of the element, respectively. The square of the flux density can be calculated as

$$B^2 = \left(B_x^2 + B_y^2\right) = \left(\frac{A_i b_i + A_j b_j + A_k b_k}{2\Delta}\right)^2 + \left(\frac{A_i c_i + A_j c_j + A_k c_k}{2\Delta}\right)^2 \tag{5-29}$$

Substituting

$$W_{mag,e} = \frac{1}{2\mu_e} l_a \mathbf{a}_e^{\mathrm{T}} \mathbf{k}_e \mathbf{a}_e \tag{5-30}$$

Then the total energy of all the elements can be calculated as

$$W_{mag} = \frac{1}{2} p_f l_a \mathbf{A}^{\mathrm{T}} \mathbf{K} \mathbf{A} = \frac{1}{2} p_f l_a \sum_{e=1}^{n} \frac{\mathbf{a}_e^{\mathrm{T}} \mathbf{k}_e \mathbf{a}_e}{\mu_e} \tag{5-31}$$

where p_f is the periodical factor when symmetrical boundaries are introduced to reduce the solution domain.

5.3.1 Flux Linkage Calculation with Magnetic Vector Potential

The flux and flux linkage of a single winding coil can be calculated directly by the magnetic vector potential difference between its two sides, which can be expressed as

$$\phi_c = A_{s1} - A_{s2}$$
$$\psi_c = N_{ct}\left(A_{s1} - A_{s2}\right)$$

(5-32)

where A_{s1} and A_{s2} are the potentials of the two sides, N_{ct} is the number of turns of the coil, ϕ_c is the flux linkage of the single turn, and ψ_c is the total flux linkage.

Then the flux and flux linkage of each phase, in which each coil is in series, can be calculated as

$$\phi_k = \sum{}' \phi_c$$
$$\psi_k = \sum \psi_c$$

(5-33)

5.3.2 Flux Linkage and Apparent Inductance Calculation with Magnetic Field Energy

In electrical machines, the energy of the magnetic field can also be written as

$$W_{mag} = \frac{1}{2}\sum_k i_k \psi_k$$

(5-34)

where i_k is the phase current and ψ_k is the phase flux linkage. Since the research in this work is focused on the three-phase permanent magnet synchronous motor, the flux linkage in the static three-phase coordinates can be expressed as

$$\begin{bmatrix} \psi_1 \\ \psi_2 \\ \psi_3 \end{bmatrix} = \begin{bmatrix} L_{11} & L_{12} & L_{13} \\ L_{21} & L_{22} & L_{23} \\ L_{31} & L_{32} & L_{33} \end{bmatrix} \cdot \begin{bmatrix} i_1 \\ i_2 \\ i_3 \end{bmatrix} + \begin{bmatrix} \psi_{1f} \\ \psi_{2f} \\ \psi_{3f} \end{bmatrix}$$

(5-35)

where ψ_{kf} (k = 1, 2, 3) are the flux linkage components due to the permanent magnets (PMs), and L_{jk} (j and k = 1, 2, 3) the self (if $j=k$) or mutual (if $j \neq k$) inductance of phase windings 1, 2, and 3, respectively.

The frozen permeability method is applied for the inductance calculation. Firstly, the reluctivity of each element is stored after the nonlinear calculation by the Newton Raphson method. Then the reluctivity is utilized for the following performance calculation and nonlinear computation is not required. The self-inductance and mutual inductance of phases are calculated while the remanence of the magnets is set to be zero. Taking the calculation of phase one and two as an example, with the stored reluctivity, when we add the phase current $i_1 = 1$, while the current of other phases are set as zeros, the self-inductance of phase one can be calculated as

$$L_{11} = 2W_{\text{mag}}^{\text{p1}} \tag{5-36}$$

where the $W_{\text{mag}}^{\text{p1}}$ is the magnetic field energy at the current status.

Similarly, when only $i_2 = 1$ A, the self-inductance of phase two can be obtained

$$L_{22} = 2W_{\text{mag}}^{\text{p2}} \tag{5-37}$$

When the excitation currents of two phases are added, i.e. $i_1 = 1$, $i_2 = 1$, the magnetic field energy at the condition can be written as $W_{\text{mag}}^{\text{p12}}$. The mutual energy can be expressed as

$$W_{\text{mag}}^{\text{mu}12} = W_{\text{mag}}^{\text{p12}} - W_{\text{mag}}^{\text{p1}} - W_{\text{mag}}^{\text{p2}} \tag{5-38}$$

Then the value of mutual inductance is exact the mutual energy, i.e.

$$L_{12} = W_{\text{mag}}^{\text{mu}12} \tag{5-39}$$

The above method for calculating inductances can be applied repetitively for different phases until all self and mutual inductances are obtained. The magnet flux linkage is calculated based on the mutual energy between phase current excitation and PMs. The field energy generated by only the magnets can be calculated as $W_{\text{mag}}^{\text{m}}$ by the linear process mentioned above when the PM remanence is restored while no current excitation is applied. Together with the PM excitation, when phase 1 is supplied with unit current and the total energy can be computed as $W_{\text{mag}}^{\text{p1m}}$. Consequently, the magnet flux linkage of phase one equals exactly the mutual energy between the magnet and phase one, i.e.

$$\psi_{1f} = W_{\text{mag}}^{\text{mulm}} = W_{\text{mag}}^{\text{plm}} - W_{\text{mag}}^{\text{pl}} - W_{\text{mag}}^{\text{m}} \tag{5-40}$$

The same steps can be conducted to obtain the permanent magnet flux linkage of other phases. Then all the expression of the inductance or flux linkage can be achieved with the calculation based on the magnetic field energy.

5.3.3 Flux Linkage and Inductance in Rotational Coordinate System

The dq0 transformation has been widely used in simplifying the steady-state performance and control of three-phase electrical machines. The transformation is the product of the Clarke transformation and the Park transformation, which can transfer the AC waveforms in a static coordinate to DC signals in a rotational reference frame. In the finite element analysis of electrical machines, the flux linkage and current in the rotational reference frame can be used in the average torque calculation. By the Clark and Park transformations, the flux linkages in the synchronous rotating reference frame can be expressed as

$$\begin{bmatrix} \psi_{\text{d}} \\ \psi_{\text{q}} \end{bmatrix} = \frac{2}{3} \begin{bmatrix} \cos\theta & \cos(\theta - \frac{2}{3}\pi) & \cos(\theta + \frac{2}{3}\pi) \\ -\sin\theta & -\sin(\theta - \frac{2}{3}\pi) & -\sin(\theta + \frac{2}{3}\pi) \end{bmatrix} \begin{bmatrix} \psi_1 \\ \psi_2 \\ \psi_3 \end{bmatrix} \tag{5-41}$$

$$\begin{bmatrix} i_{\text{d}} \\ i_{\text{q}} \end{bmatrix} = \frac{2}{3} \begin{bmatrix} \cos\theta & \cos(\theta - \frac{2}{3}\pi) & \cos(\theta + \frac{2}{3}\pi) \\ -\sin\theta & -\sin(\theta - \frac{2}{3}\pi) & -\sin(\theta + \frac{2}{3}\pi) \end{bmatrix} \begin{bmatrix} i_1 \\ i_2 \\ i_3 \end{bmatrix} \tag{5-42}$$

where ψ_{d} and ψ_{q} are the flux linkages in d and q directions in the rotating coordinate system, respectively, i_{d} is the d axis current and i_{q} is the q axis current, and θ instantaneous electrical angle. Then the flux linkage in the rotating coordinate system can be expressed as

$$\begin{aligned} \psi_{\text{d}} &= \psi_{\text{mf}} + L_{\text{d}} i_{\text{d}} \\ \psi_{\text{q}} &= L_{\text{q}} i_{\text{q}} \end{aligned} \tag{5-43}$$

where ψ_{mf} is the permanent magnet flux linkage component, L_d and L_q are the inductance in d and q axes, respectively.

By applying the frozen permeability method, the apparent inductance values of L_d and L_q can be calculated from the magnetic energy when the remanence of the permanent magnet is set as zero. Let $i_\mathrm{d} = 1$ A, and $i_\mathrm{q} = 0$ A, and we obtain

$$W_{\mathrm{mag}}^{\mathrm{d}} = \frac{3}{2}(\frac{1}{2}L_\mathrm{d}i_\mathrm{d}^2) \tag{5-44}$$

$$L_\mathrm{d} = \frac{4W_{\mathrm{mag}}^{\mathrm{d}}}{3} \tag{5-45}$$

Similarly, when $i_\mathrm{d} = 0$, $i_\mathrm{q} = 1$ A, we can obtain

$$L_\mathrm{q} = \frac{2}{3}\frac{2W_{\mathrm{mag}}^{\mathrm{q}}}{i_\mathrm{q}^2} = \frac{4W_{\mathrm{mag}}^{\mathrm{q}}}{3} \tag{5-46}$$

When the i_d and i_q are set as zero and the remanence of the permanent magnet is restored, the magnetic energy $W_{\mathrm{mag}}^{\mathrm{m}}$ can be calculated, and when the $i_\mathrm{d} = 1$, $i_\mathrm{q} = 0$, the total magnetic energy $W_{\mathrm{mag}}^{\mathrm{dm}}$ can be calculated and its relation with the magnetic field energy calculated above can be expressed as

$$W_{\mathrm{mag}}^{\mathrm{dm}} = W_{\mathrm{mag}}^{\mathrm{m}} + W_{\mathrm{mag}}^{\mathrm{d}} + W_{\mathrm{mag}}^{\mathrm{mu}} \tag{5-47}$$

where $W_{\mathrm{mag}}^{\mathrm{mu}}$ is the mutual energy between PM and the d axis current. Then the magnetic flux linkage can be calculated as

$$\psi_{\mathrm{mf}} = \frac{2}{3}\frac{W_{\mathrm{mag}}^{\mathrm{mu}}}{i_\mathrm{d}} = \frac{2W_{\mathrm{mag}}^{\mathrm{mu}}}{3} \tag{5-48}$$

5.3.4 Torque Calculation

Various calculation approaches have been developed for efficient computation of the torque profile, such as the Maxwell stress tensor method, Coulomb virtual work method and nodal force method, etc., in which the former two methods are widely used. In the topology optimization, we need to calculate the sensitivity of performance with respect to the design variable for the optimization algorithm to decide the element's material. By applying these two methods, the sensitivity can hardly be calculated directly or requiring the introduction of an adjoint vector (for the Maxwell tensors method). Therefore, to achieve an explicit expression of the torque for sensitivity analysis by the field circuit coupling method, the computation approaches based on the energy conversion and Fourier analysis are derived. Since the current excitation is applied for the finite element analysis and torque calculation, the formulation is also derived based on the co-energy, which takes the phase current and mechanical rotation degree as independent variables.

5.3.4.1 Maxwell Stress Tensor Method

Maxwell stress tensor method can be applied easily to achieve the electromagnetic torque. Firstly, select a path in the air gap that surrounds the rotor and divide by the elements into line segments. The tangential and normal components (B_t and B_n) of the flux density of the line segment can be obtained by the achieved flux density in each element. Then the torque can be calculated as

$$T = r \oint \frac{B_n B_t}{\mu_0} \mathrm{d}l \tag{5-49}$$

where r is the radius of the path.

5.3.4.2 The Virtual Work Method

According to the conservation of energy, the input electrical energy should be equal to the sum of increment of magnetic energy and output mechanical energy

$$\mathrm{d}W_{elec} - \mathrm{d}W_{mag} = \mathrm{d}W_{mech} \tag{5-50}$$

According to the Faraday's electromagnetic induction law, the change of the flux linkage leads to the induced electromotive force

$$e_k = -\frac{\mathrm{d}\psi_k}{\mathrm{d}t} \tag{5-51}$$

If the current of the kth phase is i_k, the total input electrical energy can be expressed as

$$\mathrm{d}W_{\mathrm{elec}} = \sum_{k=1}^{n} -e_k i_k \, \mathrm{d}t = \sum_{k=1}^{n} i_k \, \mathrm{d}\psi_k \tag{5-52}$$

If the displacement of the rotor is $\mathrm{d}\theta_{\mathrm{mech}}$, and the electromagnetic torque is T, the output mechanical energy can be expressed as

$$\mathrm{d}W_{\mathrm{mech}} = -T\mathrm{d}\theta_{\mathrm{mech}} \tag{5-53}$$

The conservation of the energy can be rewritten as

$$\mathrm{d}W_{\mathrm{mag}} = \sum_{k=1}^{n} i_k \, \mathrm{d}\psi_k - T\mathrm{d}\theta_{\mathrm{mech}} \tag{5-54}$$

substituting $\mathrm{d}W_{\mathrm{mag}}$ with the partial derivative

$$\mathrm{d}W_{\mathrm{mag}} = \sum_{k=1}^{n} \frac{\partial W_{\mathrm{mag}}}{\partial \psi_k} \mathrm{d}\psi_k + \frac{\partial W_{\mathrm{mag}}}{\partial \theta_{\mathrm{mech}}} \mathrm{d}\theta_{\mathrm{mech}} \tag{5-55}$$

When the flux linkage is constant, the torque can be expressed as

$$T = -\frac{\partial W_{\mathrm{mag}}}{\partial \theta_{\mathrm{mech}}} = -n_{\mathrm{p}} \frac{\partial W_{\mathrm{mag}}}{\partial \theta_{\mathrm{elec}}} \tag{5-56}$$

which can be calculated as

$$T = \frac{1}{2} p_{\mathrm{f}} l_{\mathrm{a}} \mathbf{A}^{\mathbf{T}} \frac{\partial \mathbf{K}}{\partial \theta_{\mathrm{mech}}} \mathbf{A} \tag{5-57}$$

while the permeability of each element achieved by the Newton Raphson method is kept constant.

5.3.4.3 Field Circuit Coupling Method I

Based on the energy conversion theory of electrical machines as (5-55), the output torque can be expressed as

$$T = \sum_{k=1}^{n} i_k \frac{\mathrm{d}\psi_k}{\mathrm{d}\theta_{\mathrm{mech}}} - \frac{\mathrm{d}W_{\mathrm{mag}}}{\mathrm{d}\theta_{\mathrm{mech}}} \tag{5-58}$$

The relation of the electrical degree and mechanical degree is

$$d\theta_{\text{elec}} = n_{\text{p}}d\theta_{\text{mech}} \tag{5-59}$$

where n_{p} is the number of pole pairs. If the electrical degree is applied, the (5-58) can be rewritten as

$$T\left(\theta_{\text{elec}}\right) = n_{\text{p}}\sum_{k=1}^{n} i_{k}\left(\theta_{\text{elec}}\right)\frac{d\psi_{k}\left(\theta_{\text{elec}}\right)}{d\theta_{\text{elec}}} - n_{\text{p}}\frac{dW_{\text{mag}}\left(\theta_{\text{elec}}\right)}{d\theta_{\text{elec}}} \tag{5-60}$$

The harmonics of the flux linkage can be obtained by applying the discrete Fourier transformation (DFT) when the values of flux linkage of each phase at N rotor positions are achieved. The flux linkage then can be expressed as

$$\psi_{k}\left(\theta_{\text{elec}}\right) = \frac{\psi_{k,0}^{A}}{2} + \sum_{h=1}^{M-1}\left[\psi_{k,h}^{A}\cos\left(h\theta_{\text{elec}}\right) + \psi_{k,h}^{B}\sin\left(h\theta_{\text{elec}}\right)\right] \tag{5-61}$$

If N is an even number, $N = 2M$, where the coefficients can be calculated by

$$\psi_{k,h}^{A} = \frac{2}{N}\sum_{i=0}^{N-1}\psi_{k}\left(\theta_{\text{elec},i}\right)\cos\left(h\theta_{\text{elec},i}\right) \quad \theta_{\text{elec},i} = \frac{2\pi i}{N} \ (i = 0,...,N-1) \tag{5-62}$$

$$\psi_{k,h}^{B} = \frac{2}{N}\sum_{i=0}^{N-1}\psi_{k}\left(\theta_{\text{elec},i}\right)\sin\left(h\theta_{\text{elec},i}\right) \quad \theta_{\text{elec},i} = \frac{2\pi i}{N} \ (i = 0,...,N-1) \tag{5-63}$$

With the decoupled harmonics, the flux linkage derivative with respect to the electrical degree can be computed as

$$\frac{d\psi\left(\theta_{\text{elec}}\right)}{d\theta_{\text{elec}}} = \sum_{h=0}^{M-1}\left[-hA_{h}\sin\left(h\theta_{\text{elec}}\right) + hB_{h}\cos\left(h\theta_{\text{elec}}\right)\right] \tag{5-64}$$

We write the electrical degree, flux linkage, and harmonic orders with a vector, which can be expressed as

$$\boldsymbol{\theta} = \left[\theta_{\text{elec},0}, \theta_{\text{elec},1},...\theta_{\text{elec},N\text{-}1}\right] \tag{5-65}$$

$$\boldsymbol{\Psi}_{k} = \left[\psi_{k}\left(\theta_{\text{elec},0}\right), \psi_{k}\left(\theta_{\text{elec},1}\right),...,\psi_{k}\left(\theta_{\text{elec},N-1}\right)\right] \tag{5-66}$$

$$\mathbf{M} = \begin{bmatrix} 0,1,\cdots,M-1 \end{bmatrix} \tag{5-67}$$

The coefficient vector of the flux linkage harmonics then can be written as

$$\psi_{k,h}^{A} = \frac{2}{N}\psi_{k}\cos\left(\mathbf{\theta}^{T}\mathbf{M}\right) \tag{5-68}$$

$$\psi_{k,h}^{B} = \frac{2}{N}\psi_{k}\sin\left(\mathbf{\theta}^{T}\mathbf{M}\right) \tag{5-69}$$

Then the magnetic field energy harmonic vector can also be calculated as

$$\mathbf{WA}_{h}^{mag} = \frac{2}{N}\mathbf{W}_{mag}\cos\left(\mathbf{\theta}^{T}\mathbf{M}\right) \tag{5-70}$$

$$\mathbf{WB}_{h}^{mag} = \frac{2}{N}\mathbf{W}_{mag}\sin\left(\mathbf{\theta}^{T}\mathbf{M}\right) \tag{5-71}$$

where the magnetic field energy vector is

$$\mathbf{W}_{mag} = \begin{bmatrix} W_{mag}\left(\theta_{elec,0}\right), W_{mag}\left(\theta_{elec,1}\right),..., W_{mag}\left(\theta_{elec,N-1}\right) \end{bmatrix} \tag{5-72}$$

Then the torque according to the electrical degree can be calculated as

$$\begin{aligned}
T\left(\theta_{ele}\right) = \sum_{k=1}^{n} i_{k}\left(\theta_{elec}\right)&\left[\psi_{k,h}^{B}\odot\mathbf{M}\cos\left(\mathbf{M}^{T}\theta_{elec}\right) - \psi_{k,h}^{A}\odot\mathbf{M}\sin\left(\mathbf{M}^{T}\theta_{elec}\right) \right] \\
&- \left[\mathbf{WB}_{h}^{mag}\odot\mathbf{M}\cos\left(\mathbf{M}^{T}\theta_{elec}\right) - \mathbf{WA}_{h}^{mag}\odot\mathbf{M}\sin\left(\mathbf{M}^{T}\theta_{elec}\right) \right]
\end{aligned} \tag{5-73}$$

where $\odot$ means Hadamard product, i_k is current of the kth phase.

5.3.4.4 Field Circuit Coupling Method II

As mentioned above, the torque is calculated as the flux linkage and the mechanical or electrical degree are the independent variables of the system. For a quick calculation, phase current excitation is applied directly as the independent variables. The applcation of the co-energy which takes the phase current and electrical degree as independent variables also can be applied for flexible and efficient torque calculation and sensitivity analysis in the following steps of the research.

The relation of magnetic field energy and co-energy can be expressed as

$$W_{\text{mag}} + W_{\text{Cmag}} = \sum_{k=1}^{n} \psi_k i_k \tag{5-74}$$

Then the total differential of the co-energy can be written as

$$\mathrm{d}W_{\text{Cmag}} = \sum_{k=1}^{n} \psi_k \, \mathrm{d}i_k + T \mathrm{d}\theta_{\text{mech}} \tag{5-75}$$

Similarly, according to the proportional relation of the electrical degree and mechanical degree, the torque can be calculated as

$$T = n_{\text{p}} \frac{\mathrm{d}W_{\text{Cmag}}}{\mathrm{d}\theta_{\text{elec}}} - n_{\text{p}} \sum_{k=1}^{n} \psi_k \frac{\mathrm{d}i_k}{\mathrm{d}\theta_{\text{elec}}} \tag{5-76}$$

In the frequency domain magneto-static field calculation, the sinusoidal current is usually applied as excitation. The phase current can be expressed as

$$i_k\left(\theta_{\text{elec}}\right) = I_k \sin\left(\theta_{\text{elec}} + \theta_k^{\text{ip}}\right) \tag{5-77}$$

The derivative of the phase current is

$$i_k'\left(\theta_{\text{elec}}\right) = I_k \cos\left(\theta_{\text{elec}} + \theta_k^{\text{ip}}\right) \tag{5-78}$$

Similarly, the harmonic coefficients of co-energy calculated with DFT can be expressed as

$$\mathbf{WA}_{\text{h}}^{\text{Cmag}} = \frac{2}{N} \mathbf{W}_{\text{Cmag}} \cos\left(\boldsymbol{\theta}^{\text{T}}\mathbf{M}\right) \tag{5-79}$$

$$\mathbf{WB}_{\text{h}}^{\text{Cmag}} = \frac{2}{N} \mathbf{W}_{\text{Cmag}} \sin\left(\boldsymbol{\theta}^{\text{T}}\mathbf{M}\right) \tag{5-80}$$

where the co-energy vector is

$$\mathbf{W}_{\text{mag}} = \left[W_{\text{Cmag}}\left(\theta_{\text{ele},0}\right), W_{\text{Cmag}}\left(\theta_{\text{ele},1}\right), \ldots, W_{\text{Cmag}}\left(\theta_{\text{ele},N-1}\right)\right] \tag{5-81}$$

Then the torque can be calculated as

$$T\left(\theta_{\text{elec}}\right) = \left[\mathbf{WB}_{\text{h}}^{\text{Cmag}} \odot \mathbf{M}\cos\left(\mathbf{M}^{\text{T}}\theta_{\text{elec}}\right) - \mathbf{WA}_{\text{h}}^{\text{Cmag}} \odot \mathbf{M}\sin\left(\mathbf{M}^{\text{T}}\theta_{\text{elec}}\right) \right]$$
$$-\sum_{k=1}^{n} i_{k}'\left(\theta_{\text{elec}}\right)\left[\boldsymbol{\psi}_{k,h}^{\text{A}} \odot \cos\left(\mathbf{M}^{\text{T}}\theta_{\text{elec}}\right) + \boldsymbol{\psi}_{k,h}^{\text{B}} \odot \sin\left(\mathbf{M}^{\text{T}}\theta_{\text{elec}}\right) \right] \tag{5-82}$$

5.3.4.5 Torque Calculation Based on dq0 Transformation

According to the field circuit coupling method, the torque can be expressed as

$$T = 1.5 n_{\text{p}} (\psi_{\text{d}} i_{\text{q}} - \psi_{\text{q}} i_{\text{d}}) \tag{5-83}$$

Different from the torque calculation methods mentioned before, this method only considers and calculates the torque from the input side of the energy, i.e. the electrical power. Since it does not consider the magnetic field energy variation, this method cannot calculate the torque at different rotor positions accurately. However, in the steady-state operation, the magnetic field energy at a certain rotor position is constant after each revolution, which means all the input electrical power is transferred into mechanical power. Therefore, this method can be used for calculating the average torque.

5.4 Topology Optimization with the Density Method

The aim of the topology optimization method is the optimal layout of materials in the design domain. When the FEM is applied for performance analysis, the design domain is divided into finite elements. This research is focused on the topology optimization of the soft magnetic material part of the electrical machines. If we define the existence of the material in the element as the variable, namely each element can only switch between 1 and 0 which represents the material is soft magnetic material or air. The magnetic property i.e. the permeability (or magnetic reluctivity) is then changed according to the value of the design variable. This problem belongs to the optimization of discrete variables. The ON-OFF methods require multiple times of the finite element computation so the whole optimization can be time-consuming. Since it is hard to run through all the possible solutions for the problem of the optimal layout with reduced material, this method relies highly on the searching ability of the algorithms. For example, for a problem with N design variables, if only M elements are required to be arranged with solid materials, the number of solutions is

$$C_N^M = \frac{N!}{(N-M)!M!} \tag{5-84}$$

The density method aims to decrease the optimization complexity by change the discrete problem into a continuous problem. The design variables named as element density are defined as continuous (the interval is usually defined as (0-1]) for the sensitivity analysis. Then the iteration of the algorithm can be based on the achieved sensitivity for quick convergence.

To take advantage of the density method, this section presents the building of the interpolation function, the sensitivity analysis of the performances, optimization algorithms the framework establishment, and the postprocessing method for the topology optimization of ferromagnetic component PM electrical machines.

5.4.1 Interpolation method

For electrical machines analysis, the material properties usually contain permeability, permittivity, conductivity, etc. The classification of materials is based on the difference of the properties. Accordingly, to transit one material to another continuously, the interpolation function should be used for the property parameters. The electromagnetic analysis of PM electrical machines usually takes account of four kinds of materials including conductors, air, soft magnetic materials, and PM. Some research has been done for the multi-material optimization of electrical devices which proves the possibility of the topology optimization considering many materials at the same time. However, the optimization can be very complicated, and the strategy requires further improvement. This study focuses on the topology optimization of the soft magnetic material part. Therefore, single property interpolation of the reluctivity or permeability is utilized. The commonly applied formulations of the interpolation are expressed as below and defined as method 1 to 3

$$\mu_e = \mu_{air} + \left(\mu_{fe}(B) - \mu_{air} \right) \rho_e^p \tag{5-85}$$

$$\mu_e = \mu r_{air} \mu r_{fe}^{\rho_e}(B) \mu_0 \tag{5-86}$$

$$\frac{1}{\mu_e} = \frac{1}{\mu_{\text{fe}}}(B_e)\rho_e^p + \frac{1}{\mu_{\text{air}}}(1-\rho_e^p)$$

(5-87)

where μ_{air} μ_{fe} are the permeability of the air and ferromagnetic material respectively, ρ_e is the element density of element e and p is the penalization power. Taking the linear property as an example, when we set the relative permeability of air and steel as 1 and 3000, the interpolation permeability with respect to the density can be illustrated as Figure 5-9 according to the equation (5-85)-(5-87).

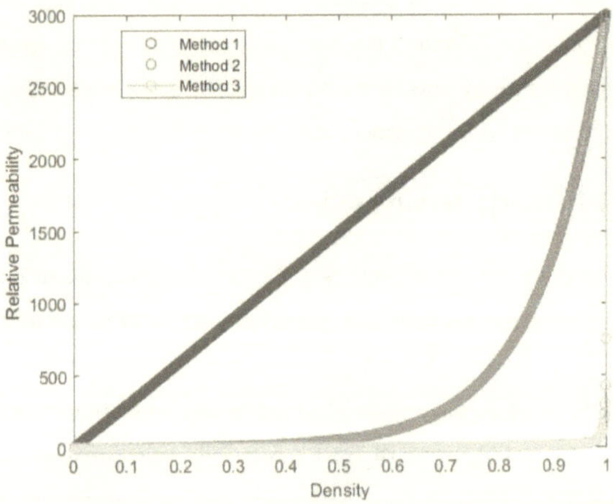

Figure 5-9 Relative permeability variation with the density of design interpolation method

In this research, the third method of interpolation is applied. The interpolation of inverse of relative permeability with respect to the density and p is illustrated in Figure 5-10. Particularly, considering the impact of saturation and nonlinear solution, the properties of the soft magnetic materials are calculated according to the BH curve. The minimum reluctivity at the BH curve is defined as the initial reluctivity of the soft magnetic material defined. Simultaneously, the interpolation will also be applied to other points at the BH curve for the elements in the design domain. When the nonlinear solution of the finite element analysis is achieved at each iteration step, the $v_{\text{steel}}(B)$ is calculated with the interpolation function and used for the sensitivity calculation for the next iteration.

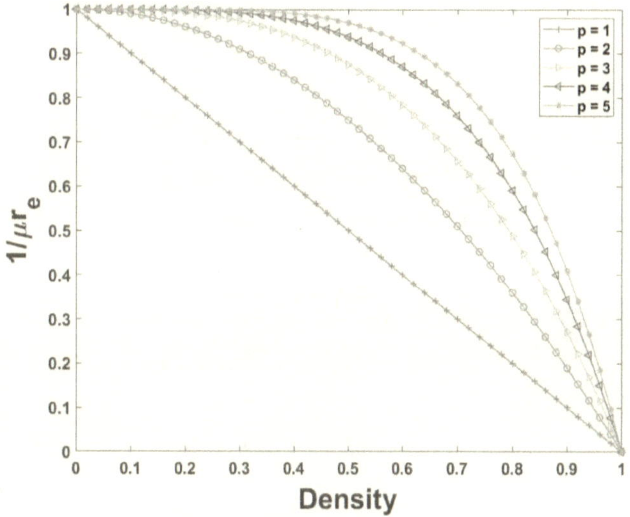

Figure 5-10 $1/\mu r_e$ variation with respect to the element density of different p.

5.4.2 Sensitivity Analysis of Field Energy

As introduced in the performance calculation section, the performance of the electrical machine can be derived from the energy of the system. Based on this feature, the sensitivity analysis of the performance with respect to the design variables is proposed to be bridged by the element energy. According that the partial derivative of the **F** is zero with respect to the density because **F** is related to the excitation,

$$\frac{\partial(\mathbf{KA})}{\partial\rho_e} = \frac{\partial\mathbf{F}}{\partial\rho_e} = 0 \tag{5-88}$$

Then

$$\mathbf{K}\frac{\partial\mathbf{A}}{\partial\rho_e} = -\frac{\partial\mathbf{K}}{\partial\rho_e}\mathbf{A} \tag{5-89}$$

Simultaneously, according to the symmetry of the stiffness matrix

$$\left(\mathbf{K}\frac{\partial \mathbf{A}}{\partial \rho_{\mathrm{e}}}\right)^{\mathrm{T}} = \left(-\frac{\partial \mathbf{K}}{\partial \rho_{\mathrm{e}}}\mathbf{A}\right)^{\mathrm{T}} \tag{5-90}$$

$$\frac{\partial \mathbf{A}^{\mathrm{T}}}{\partial \rho_{\mathrm{e}}}\mathbf{K} = -\mathbf{A}^{\mathrm{T}}\frac{\partial \mathbf{K}^{\mathrm{T}}}{\partial \rho_{\mathrm{e}}} = -\mathbf{A}^{\mathrm{T}}\frac{\partial \mathbf{K}}{\partial \rho_{\mathrm{e}}} \tag{5-91}$$

Based on the derivation above, we can achieve

$$\begin{aligned}
\frac{\partial W_{\mathrm{mag}}}{\partial \rho_{\mathrm{e}}} &= \frac{1}{2}p_{\mathrm{f}}l_{\mathrm{a}}\frac{\partial \mathbf{A}^{\mathrm{T}}\mathbf{K}\mathbf{A}}{\partial \rho_{\mathrm{e}}} \\
&= \frac{1}{2}p_{\mathrm{f}}l_{\mathrm{a}}\left(\frac{\partial \mathbf{A}^{\mathrm{T}}}{\partial \rho_{\mathrm{e}}}\mathbf{K}\mathbf{A} + \mathbf{A}^{\mathrm{T}}\frac{\partial \mathbf{K}}{\partial \rho_{\mathrm{e}}}\mathbf{A} + \mathbf{A}^{\mathrm{T}}\mathbf{K}\frac{\partial \mathbf{A}}{\partial \rho_{\mathrm{e}}}\right) \\
&= \frac{1}{2}p_{\mathrm{f}}l_{\mathrm{a}}\left(-\mathbf{A}^{\mathrm{T}}\frac{\partial \mathbf{K}}{\partial \rho_{\mathrm{e}}}\mathbf{A} + \mathbf{A}^{\mathrm{T}}\frac{\partial \mathbf{K}}{\partial \rho_{\mathrm{e}}}\mathbf{A} - \mathbf{A}^{\mathrm{T}}\frac{\partial \mathbf{K}}{\partial \rho_{\mathrm{e}}}\mathbf{A}\right) \\
&= -\frac{1}{2}p_{\mathrm{f}}l_{\mathrm{a}}\mathbf{A}^{\mathrm{T}}\frac{\partial \mathbf{K}}{\partial \rho_{\mathrm{e}}}\mathbf{A} \\
&= -\frac{1}{2}p_{\mathrm{f}}l_{\mathrm{a}}p\left[\frac{1}{\mu_{\mathrm{fe}}}(B_{\mathrm{e}}) - \frac{1}{\mu_{\mathrm{air}}}\right]\rho_{\mathrm{e}}^{p-1}\mathbf{a}_{\mathrm{e}}^{\mathrm{T}}\mathbf{k}_{\mathrm{e}}\mathbf{a}_{\mathrm{e}}
\end{aligned} \tag{5-92}$$

This equation about the sensitivity of magnetic field energy with respect to the element density offers the precondition for the following sensitivity derivation of the performances.

5.4.3 Sensitivity Analysis of Flux Linkage and Inductance

The flux linkage of each phase is composed of the self and mutual components. Taking the flux linkage of one phase in the equation as an example, we can calculate its sensitivity as

$$\frac{\partial \psi_1}{\partial \rho_{\mathrm{e}}} = i_1\frac{\partial L_{11}}{\partial \rho_{\mathrm{e}}} + i_2\frac{\partial L_{12}}{\partial \rho_{\mathrm{e}}} + i_3\frac{\partial L_{13}}{\partial \rho_{\mathrm{e}}} + \frac{\partial \psi_{1\mathrm{f}}}{\partial \rho_{\mathrm{e}}} \tag{5-93}$$

$$\begin{aligned}
\frac{\partial L_{11}}{\partial \rho_{\mathrm{e}}} &= 2\frac{\partial\left(W_{\mathrm{mag}}^{\mathrm{p1}}\right)}{\partial \rho_{\mathrm{e}}} = p_{\mathrm{f}}l_{\mathrm{a}}\frac{\partial\left[\left(\mathbf{A}_{\mathrm{mag}}^{\mathrm{p1}}\right)^{\mathrm{T}}\mathbf{K}_{\mathrm{mag}}^{\mathrm{p1}}\mathbf{A}_{\mathrm{mag}}^{\mathrm{p1}}\right]}{\partial \rho_{\mathrm{e}}} \\
&= -p_{\mathrm{f}}l_{\mathrm{a}}p\left[\frac{1}{\mu_{\mathrm{fe}}}(B_{\mathrm{e}}) - \frac{1}{\mu_{\mathrm{air}}}\right]\rho_{\mathrm{e}}^{p-1}\left[\left(\mathbf{a}_{\mathrm{mag,e}}^{\mathrm{p1}}\right)^{\mathrm{T}}\mathbf{k}_{\mathrm{mag,e}}^{\mathrm{p1}}\mathbf{a}_{\mathrm{mag,e}}^{\mathrm{p1}}\right]
\end{aligned} \tag{5-94}$$

$$\frac{\partial L_{12}}{\partial \rho_{\mathrm{e}}} = \frac{\partial \left(W_{\mathrm{mag}}^{\mathrm{mu12}} \right)}{\partial \rho_{\mathrm{e}}} = \frac{1}{2} p_{\mathrm{f}} l_{\mathrm{a}} \frac{\partial \left[\left(\mathbf{A}_{\mathrm{mag}}^{\mathrm{mu12}} \right)^{\mathbf{T}} \mathbf{K}_{\mathrm{mag}}^{\mathrm{mu12}} \mathbf{A}_{\mathrm{mag}}^{\mathrm{mu12}} \right]}{\partial \rho_{\mathrm{e}}} \tag{5-95}$$

$$= -\frac{1}{2} p_{\mathrm{f}} l_{\mathrm{a}} p \left[\frac{1}{\mu_{\mathrm{fe}}} (B_{\mathrm{e}}) - \frac{1}{\mu_{\mathrm{air}}} \right] \rho_{\mathrm{e}}^{p-1} \left[\left(\mathbf{a}_{\mathrm{mag,e}}^{\mathrm{mu12}} \right)^{\mathbf{T}} \mathbf{k}_{\mathrm{mag,e}}^{\mathrm{mu12}} \mathbf{a}_{\mathrm{mag,e}}^{\mathrm{mu12}} \right]$$

$$\frac{\partial L_{13}}{\partial \rho_{\mathrm{e}}} = -\frac{1}{2} p_{\mathrm{f}} l_{\mathrm{a}} p \left[\frac{1}{\mu_{\mathrm{fe}}} (B_{\mathrm{e}}) - \frac{1}{\mu_{\mathrm{air}}} \right] \rho_{\mathrm{e}}^{p-1} \left[\left(\mathbf{a}_{\mathrm{mag,e}}^{\mathrm{mu13}} \right)^{\mathbf{T}} \mathbf{k}_{\mathrm{mag,e}}^{\mathrm{mu13}} \mathbf{a}_{\mathrm{mag,e}}^{\mathrm{mu13}} \right] \tag{5-96}$$

$$\frac{\partial \Psi_{\mathrm{1f}}}{\partial \rho_{\mathrm{e}}} = -\frac{1}{2} p_{\mathrm{f}} l_{\mathrm{a}} p \left[\frac{1}{\mu_{\mathrm{fe}}} (B_{\mathrm{e}}) - \frac{1}{\mu_{\mathrm{air}}} \right] \rho_{\mathrm{e}}^{p-1} \left[\left(\mathbf{a}_{\mathrm{mag,e}}^{\mathrm{mulm}} \right)^{\mathbf{T}} \mathbf{k}_{\mathrm{mag,e}}^{\mathrm{mulm}} \mathbf{a}_{\mathrm{mag,e}}^{\mathrm{mulm}} \right] \tag{5-97}$$

where the vector potential and stiffness matrix element of the same superscripts with assembled ones mean their values at the current condition.

The sensitivity of inductances and flux linkages of other phases can be derived by the same process. For the model in the synchronous rotating reference frame, a similar calculation of the flux linkage sensitivity is obtained. The sensitivity of the flux linkage can be calculated as

$$\frac{\partial \Psi_{\mathrm{d}}}{\partial \rho_{\mathrm{e}}} = -\frac{1}{3} p_{\mathrm{f}} l_{\mathrm{a}} p \left[\frac{1}{\mu_{\mathrm{fe}}} (B_{\mathrm{e}}) - \frac{1}{\mu_{\mathrm{air}}} \right] \rho_{\mathrm{e}}^{p-1} \left[\left(\mathbf{a}_{\mathrm{mag,e}}^{\mathrm{mu}} \right)^{\mathbf{T}} \mathbf{k}_{\mathrm{mag,e}}^{\mathrm{mu}} \mathbf{a}_{\mathrm{mag,e}}^{\mathrm{mu}} + 2 i_{d} \left(\mathbf{a}_{\mathrm{mag,e}}^{\mathrm{d}} \right)^{\mathbf{T}} \mathbf{k}_{\mathrm{mag,e}}^{\mathrm{d}} \mathbf{a}_{\mathrm{mag,e}}^{\mathrm{d}} \right]$$

$$\tag{5-98}$$

$$\frac{\partial \Psi_{\mathrm{q}}}{\partial \rho_{\mathrm{e}}} = -\frac{2}{3} i_{q} p_{\mathrm{f}} l_{\mathrm{a}} p \left[\frac{1}{\mu_{\mathrm{fe}}} (B_{\mathrm{e}}) - \frac{1}{\mu_{\mathrm{air}}} \right] \rho_{\mathrm{e}}^{p-1} \left[\left(\mathbf{a}_{\mathrm{mag,e}}^{\mathrm{q}} \right)^{\mathbf{T}} \mathbf{k}_{\mathrm{mag,e}}^{\mathrm{q}} \mathbf{a}_{\mathrm{mag,e}}^{\mathrm{q}} \right] \tag{5-99}$$

where the element vector potential and stiffness matrix of the same superscripts with the total magnetic field energy in (5-44)-(5-48) mean their values under the current condition.

5.4.4 Sensitivity Analysis of Torque

Since the torque calculation method is based on the energy conversion by the direct sensitivity derivation, which derives the torque by the differential of input electrical energy and the coupled magnetic field energy with respect to the angular displacement in mechanical degrees. According to the sensitivity analysis of the flux linkage and co-energy, the sensitivity of torque based on the field circuit coupling method II can be expressed as

$$
\begin{aligned}
\frac{\partial T(\theta_{\text{elec}})}{\partial \rho_e} = n_p &\left[\frac{\partial \mathbf{WB}_h^{\text{Cmag}}}{\partial \rho_e} \odot \mathbf{M} \cos\left(\mathbf{M}^T \theta_{\text{elec}}\right) - \frac{\partial \mathbf{WA}_h^{\text{Cmag}}}{\partial \rho_e} \odot \mathbf{M} \sin\left(\mathbf{M}^T \theta_{\text{elec}}\right) \right] \\
- n_p \sum_{k=1}^{n} i_k'(\theta_{\text{elec}}) &\left[\frac{\partial \psi_{k,h}^A}{\partial \rho_e} \odot \cos\left(\mathbf{M}^T \theta_{\text{elec}}\right) + \frac{\partial \psi_{k,h}^B}{\partial \rho_e} \odot \sin\left(\mathbf{M}^T \theta_{\text{elec}}\right) \right]
\end{aligned}
\tag{5-100}
$$

The sensitivity of the harmonic coefficients of the co-energy and the flux linkage can be calculated as

$$
\frac{\partial \mathbf{WA}_h^{\text{Cmag}}}{\partial \rho_e} = \frac{2}{N} \frac{\partial \mathbf{W}_{\text{Cmag}}}{\partial \rho_e} \cos\left(\theta^T \mathbf{M}\right)
\tag{5-101}
$$

$$
\frac{\partial \mathbf{WB}_h^{\text{Cmag}}}{\partial \rho_e} = \frac{2}{N} \frac{\partial \mathbf{W}_{\text{Cmag}}}{\partial \rho_e} \sin\left(\theta^T \mathbf{M}\right)
\tag{5-102}
$$

where

$$
\frac{\partial W_{\text{Cmag}}}{\partial \rho_e} = \left[\frac{\partial W_{\text{Cmag}}(\theta_{\text{elec},0})}{\partial \rho_e}, \frac{\partial W_{\text{Cmag}}(\theta_{\text{elec},1})}{\partial \rho_e}, \dots, \frac{\partial W_{\text{Cmag}}(\theta_{\text{elec},N-1})}{\partial \rho_e} \right]
\tag{5-103}
$$

Similarly, the sensitivity of flux linkage can be obtained as

$$
\frac{\partial \psi_{k,h}^A}{\partial \rho_e} = \frac{2}{N} \frac{\partial \psi_k}{\partial \rho_e} \cos\left(\theta^T \mathbf{M}\right)
\tag{5-104}
$$

$$
\frac{\partial \psi_{k,h}^B}{\partial \rho_e} = \frac{2}{N} \frac{\partial \psi_k}{\partial \rho_e} \sin\left(\theta^T \mathbf{M}\right)
\tag{5-105}
$$

$$
\frac{\partial \psi_k}{\partial \rho_e} = \left| \frac{\partial \psi_k(\theta_{\text{elec},0})}{\partial \rho_e}, \frac{\partial \psi_k(\theta_{\text{elec},1})}{\partial \rho_e}, \dots, \frac{\partial \psi_k(\theta_{\text{elec},N-1})}{\partial \rho_e} \right|
\tag{5-106}
$$

The torque sensitivity calculation based on the field circuit coupling method I and the steady-state torque calculation method can be also derived in a similar way.

5.4.5 Optimization Algorithms

5.4.5.1 Optimality Criteria Method

Optimality Criteria (OC) method [1, 43] is widely used in the structural design optimization at the early stage. It is a heuristic method based on the Lagrangian function which is designed according to the design criteria. Then the iteration and updating of the design variables of the optimization are based on the function until the convergence. The Lagrangian multipliers are found through an iterative process. This algorithm has a fast convergence speed, high calculation efficiency, and is for programing so that very suitable for application in practical engineering. The Lagrangian L is established as

$$L = C + \lambda\left(V - fV_0\right) + \boldsymbol{\lambda}_1^{\mathbf{T}}\left(\mathbf{KA} - \mathbf{F}\right) + \boldsymbol{\lambda}_2^{T}\left(\mathbf{x}_{\min} - \mathbf{x}_e\right) + \boldsymbol{\lambda}_3^{T}\left(\mathbf{x}_e - \mathbf{x}_{\max}\right) \quad (5\text{-}107)$$

where C is the objective function $\lambda, \lambda_1, \lambda_2, \lambda_3$ are the Lagrangian multipliers for the constraints. The extreme value is achieved when the Kuhn-Tucker necessary condition, i.e. the derivatives of the Lagrangian function with respect to the design variables are zero.

$$\begin{cases} \dfrac{\partial L}{\partial x_e} = \dfrac{\partial C}{\partial x_e} + \lambda\dfrac{\partial V}{\partial x_e} + \boldsymbol{\lambda}_1^{T}\dfrac{\partial(\mathbf{KA})}{\partial x_e} - \lambda_2^e + \lambda_3^e = 0 \\ V = fV_0 \\ \mathbf{F} = \mathbf{KA} \\ \lambda_4^e\left(x_{\min} - x_e\right) = 0 \\ \lambda_5^e\left(x_e - x_{\min}\right) = 0 \\ \lambda_4^e > 0 \quad \lambda_5^e > 0 \quad e = 1, \cdots, N \\ x_{\min} \le x_e \le x_{\max} \end{cases} \quad (5\text{-}108)$$

The value of the Lagrangian multipliers for the variable constraints can be categorized into three cases when the extremum of function is obtained. The first case is that the lower and upper bound constraints are not active, i.e.

$$x_{\min} < x_e < x_{\max} \quad (5\text{-}109)$$

then

$$\lambda_4^e = \lambda_5^e = 0 \quad (5\text{-}110)$$

The other two cases are that the value of the design variable equals two the bounds, which is

$$\begin{cases} x_e = x_{min} \\ \lambda_4^e \geq 0 \\ \lambda_5^e = 0 \end{cases} \text{ or } \begin{cases} x_e = x_{max} \\ \lambda_4^e = 0 \\ \lambda_5^e \geq 0 \end{cases} \tag{5-111}$$

Considering the case that the variables are within the limit, we can have

$$\frac{\partial \mathbf{A}^T}{\partial \mathbf{x}_e} \mathbf{K}\mathbf{A} + \mathbf{A}^T \frac{\partial \mathbf{K}}{\partial \mathbf{x}_e} \mathbf{A} + \mathbf{A}^T \mathbf{K} \frac{\partial \mathbf{A}}{\partial \mathbf{x}_e} + \lambda \frac{\partial \mathbf{V}}{\partial \mathbf{x}_e} + \lambda_1^T \left(\frac{\partial \mathbf{K}}{\partial \mathbf{x}_e} \mathbf{A} + \mathbf{K} \frac{\partial \mathbf{A}}{\partial \mathbf{x}_e} \right) = 0 \tag{5-112}$$

According to the symmetry of the stiffness matrix, the equation can be simplified as

$$\frac{\partial A^T}{\partial \mathbf{x}_e} \left(2\mathbf{K}\mathbf{A} + \mathbf{K}\lambda_2 \right) + \lambda_1^T \frac{\partial \mathbf{K}}{\partial \mathbf{x}_e} \mathbf{A} + p\left(x_e\right)^{p-1} \left(\mathbf{a}_e\right)^T \mathbf{k}_0 \mathbf{a}_e + \lambda v_e = 0 \tag{5-113}$$

To simplify the equation further, the multiplier λ_1 can be artificially set as

$$\lambda_1 = -2\mathbf{A} \tag{5-114}$$

Then we have

$$-p\left(x_e\right)^{p-1} \left(\mathbf{a}_e\right)^T \mathbf{k}_0 \mathbf{a}_e + \lambda v_e = 0 \tag{5-115}$$

The design optimization criteria is then obtained as

$$C_e^k = \frac{p\left(x_e\right)^{p-1} \left(\mathbf{a}_e\right)^T \mathbf{k}_0 \mathbf{a}_e}{\lambda v_e} \tag{5-116}$$

Considering the boundaries of the design variable, its updating rules can be expressed as

$$x_e^{k+1} = \begin{cases} \left(C_e^k\right)^\varsigma x_e^k & x_{min} < \left(C_e^k\right)^\varsigma x_e^k < x_{max} \\ x_{min} & \left(C_e^k\right)^\varsigma x_e^k \leq x_{min} \\ x_{max} & \left(C_e^k\right)^\varsigma x_e^k \geq x_{max} \end{cases} \tag{5-117}$$

where ζ is the damping coefficient proposed to ensure the stability of the calculation.

The iteration of the design variables requires the solution of the criteria at the current step which needs the Lagrangian multiplier λ solved at first. Their solutions are sought under satisfying the volume constraint. Assuming the volume of the structure at step k+1 with the updated design variable from the kth step as V_{k+1}, then V_{k+1} still meet the constraint

$$V_{k+1} = fV_0 \tag{5-118}$$

By applying the Fourier expansion and ignoring the high order terms, we can get

$$V_{k+1} - V_k = fV_0 - V_k = \sum_{e=1}^{N} \frac{\partial V}{\partial x_e^k}\left(x_e^{k+1} - x_e^k\right) = \sum_{e=1}^{N} v_e\left(x_e^{k+1} - x_e^k\right) \tag{5-119}$$

According to the possible cases for updating the design variable of the next step, i.e.

$$\left(C_e^k\right)^{\varsigma} x_e^k \leq x_{min} \tag{5-120}$$

$$\left(C_e^k\right)^{\varsigma} x_e^k \geq x_{max} \tag{5-121}$$

$$x_{min} < \left(C_e^k\right)^{\varsigma} x_e^k < x_{max} \tag{5-122}$$

The equation can be rewritten as

$$fV_0 - V_k = \sum_A v_e\left(x_{min} - x_e^k\right) + \sum_B v_e\left(x_{max} - x_e^k\right) + \sum_C v_e\left(C_e^k x_e^k - x_e^k\right) \tag{5-123}$$

Substituting equation (5-116) to (5-123), we can obtain

$$fV_0 - V_k - \sum_A v_e\left(x_{min} - x_e^k\right) - \sum_B v_e\left(x_{max} - x_e^k\right) = \sum_C \frac{p\left(x_e\right)^{p-1}\left(\mathbf{a}_e\right)^T \mathbf{k}_e\mathbf{a}_e - \lambda v_e}{\lambda} x_e^k \tag{5-124}$$

which can be solved by dichotomy until the λ is achieved.

As introduced above, the solving of the problem by OC is based on the iteration according to the established criteria. For problems with constraints such as the volume, the gradient information for each design variable is easy to be achieved. That means the iteration and convergence can be fast. For the problem with constraints requiring complicated sensitivity computation process, the iteration computation burden may increase sharply. Therefore, OC may be utilized for the cases with few constraints with easily obtained gradient information. On the other hand, the solution of OC can hardly to be ensured as the globally optimal because the algorithm is designed for the extreme value searching. In this case, the multi-start method can be adopted for enhancing the global optimal solution ability.

5.4.5.2 The Method of Moving Asymptotes

As introduced above, the OC method can be utilized for topology optimization problems with a singular simple volume constraint. For the remaining topology optimization problem with multi-constraints or other non-volume constraints, the algorithm called Method of Moving Asymptotes (MMA) is widely applied [44, 45].

Taking the standard optimization problem as an example

$$f_0(\mathbf{x})$$
$$f_i(\mathbf{x}) \le 0, i = 1,\ldots, m \qquad (5\text{-}125)$$
$$\mathbf{x} \in X$$

where $f_1, f_1, \ldots, f_m$ are continuously differentiable functions on the design domain of X and x is the design variable vector, $X = \left\{ \mathbf{x} \in R^n \mid x_j^{\min} \le x_j \le x_j^{\max}, j = 1,\ldots, n \right\}$, where $x_j^{\min}$ and $x_j^{\max}$ are the lower and upper bounds of the design variable.

By MMA, it is transformed into an optimization problem with the form of

$$f_0(\mathbf{x}) + a_0 z + \sum_{i=1}^{m} \left(c_i y_i + \frac{1}{2} d_i y_i^2 \right)$$
$$f_i(\mathbf{x}) - a_i z - y_i \le 0, i = 1,\ldots, m \qquad (5\text{-}126)$$
$$\mathbf{x} \in X, \mathbf{y} \ge \mathbf{0}, z \ge 0$$

a_0, a_i, c_i and d_i meet the requirements of $a_0 > 0$, $a_i \ge 0$, $c_i \ge 0, d_i \ge 0$, $c_i + d_i > 0$, and, $a_i c_i > a_0$ while $y_1, \ldots, y_m$ and z are additional variables for reducing the solving difficulty.

The problem of the new form are solved by the following steps: the design objective and constraint functions $f_i(\mathbf{x})$ are firstly substituted by convex functions $\tilde{f}_i^{(k)}(\mathbf{x})$ with the variables $\left(\mathbf{x}^{(k)}, \mathbf{y}^{(k)}, z^{(k)}\right)$ of the current step. The approximate functions are based on the gradient information of the current step and the parameters $l_j^{(k)}$ and $u_j^{(k)}$. by solving the established strict convex problem, the unique optimal solution becomes the point $\left(\mathbf{x}^{(k+1)}, \mathbf{y}^{(k+1)}, z^{(k+1)}\right)$ of the next iteration. The problem with the convex functions has the form

$$
\begin{aligned}
\min: \quad & \tilde{f}_0^{(k)}(\mathbf{x}) + a_0 z + \sum_{i=1}^{m}\left(c_i y_i + \frac{1}{2} d_i y_i^2\right) \\
\text{s.t.} \quad & \tilde{f}_i^{(k)}(\mathbf{x}) - a_i z - y_i \leq 0, i = 1,\ldots,m \\
& \mathbf{x} \in X^{(k)}, \mathbf{y} \geq \mathbf{0}, z \geq 0
\end{aligned}
\tag{5-127}
$$

where the

$$
X^{(k)} = \left\{\mathbf{x} \in X \mid 0.9 l_j^{(k)} + 0.1 x_j^{(k)} \leq x_j \leq 0.9 u_j^{(k)} + 0.1 x_j^{(k)}, j = 1,\ldots,n\right\} \tag{5-128}
$$

And the convex surrogate functions are expressed as

$$
\begin{aligned}
& \tilde{f}_i^{(k)}(\mathbf{x}) = \sum_{j=1}^{n}\left(\frac{p_{ij}^{(k)}}{u_j^{(k)} - x_j} + \frac{q_{ij}^{(k)}}{x_j - l_j^{(k)}}\right) + r_i^{(k)}, i = 0,1,\ldots,m \\
& \alpha_j^{(k)} \leq x_j \leq \beta_j^{(k)}, \quad j = 1,\ldots,n \\
& y_i \geq 0, \qquad\qquad i = 1,\ldots,m \\
& z \geq 0
\end{aligned}
\tag{5-129}
$$

in which

$$
p_{ij}^{(k)} = \left(u_j^{(k)} - x_j^{(k)}\right)^2\left(1.001\left(\frac{\partial f_i}{\partial x_j}(\mathbf{x}^{(k)})\right)^{+} + 0.001\left(\frac{\partial f_i}{\partial x_j}(\mathbf{x}^{(k)})\right)^{-} + \frac{10^{-5}}{x_j^{\max} - x_j^{\min}}\right)
$$

$$
q_{ij}^{(k)} = \left(x_j^{(k)} - l_j^{(k)}\right)^2\left(0.001\left(\frac{\partial f_i}{\partial x_j}(\mathbf{x}^{(k)})\right)^{+} + 1.001\left(\frac{\partial f_i}{\partial x_j}(\mathbf{x}^{(k)})\right)^{-} + \frac{10^{-5}}{x_j^{\max} - x_j^{\min}}\right)
\tag{5-130}
$$

$$
r_i^{(k)} = f_i\left(\mathbf{x}^{(k)}\right) - \sum_{j=1}^{n}\left(\frac{p_{ij}^{(k)}}{u_j^{(k)} - x_j^{(k)}} + \frac{q_{ij}^{(k)}}{x_j^{(k)} - l_j^{(k)}}\right)
$$

where $\left(\dfrac{\partial f_i}{\partial x_j}\left(\mathbf{x}^{(k)}\right)\right)^{+}$ and $\left(\dfrac{\partial f_i}{\partial x_j}\left(\mathbf{x}^{(k)}\right)\right)^{-}$ are the maximum value achieved by comparing

$\dfrac{\partial f_i}{\partial x_j}\left(\mathbf{x}^{(k)}\right)$ and $-\dfrac{\partial f_i}{\partial x_j}\left(\mathbf{x}^{(k)}\right)$ with 0 respectively. The lower asymptotes $l_j^{(k)}$ and the upper

asymptotes $u_j^{(k)}$ of the first two iteration steps are defined as

$$
\begin{aligned}
l_j^{(k)} &= x_j^{(k)} - 0.5\left(x_j^{max} - x_j^{min}\right) \\
u_j^{(k)} &= x_j^{(k)} + 0.5\left(x_j^{max} - x_j^{min}\right)
\end{aligned}
\tag{5-131}
$$

and then the updating in the next steps follows

$$
\begin{aligned}
l_j^{(k)} &= x_j^{(k)} - \gamma_j^{(k)}\left(x_j^{(k-1)} - l_j^{(k-1)}\right) \\
u_j^{(k)} &= x_j^{(k)} + \gamma_j^{(k)}\left(u_j^{(k-1)} - x_j^{(k-1)}\right)
\end{aligned}
\tag{5-132}
$$

where

$$
\gamma_j^{(k)} = \begin{cases}
0.7 & \text{if} \quad \left(x_j^{(k)} - x_j^{(k-1)}\right)\left(x_j^{(k-1)} - x_j^{(k-2)}\right) < 0 \\
1.2 & \text{if} \quad \left(x_j^{(k)} - x_j^{(k-1)}\right)\left(x_j^{(k-1)} - x_j^{(k-2)}\right) > 0 \\
1 & \text{if} \quad \left(x_j^{(k)} - x_j^{(k-1)}\right)\left(x_j^{(k-1)} - x_j^{(k-2)}\right) = 0
\end{cases}
\tag{5-133}
$$

Meanwhile, if the asymptotes should be satisfied the bounds

$$
\begin{aligned}
l_j^{(k)} &\leq x_j^{(k)} - 0.01\left(x_j^{max} - x_j^{min}\right) \\
l_j^{(k)} &\geq x_j^{(k)} - 10\left(x_j^{max} - x_j^{min}\right) \\
u_j^{(k)} &\geq x_j^{(k)} + 0.01\left(x_j^{max} - x_j^{min}\right) \\
u_j^{(k)} &\leq x_j^{(k)} + 10\left(x_j^{max} - x_j^{min}\right)
\end{aligned}
\tag{5-134}
$$

Otherwise, they are set as the boundary values which they violate.

The first-order approximation is always used for the original functions

$$
\tilde{f}_i^{(k)}\left(\mathbf{x}^{(k)}\right) = f_i\left(\mathbf{x}^{(k)}\right) \quad \text{and} \quad \frac{\partial \tilde{f}_i^{(k)}}{\partial x_j}\left(\mathbf{x}^{(k)}\right) = \frac{\partial f_i}{\partial x_j}\left(\mathbf{x}^{(k)}\right)
\tag{5-135}
$$

A globally convergent solution can still not be guaranteed if the original problem itself is not convex. The global convergence MMA is proposed with a double loop processing of the iteration. The iteration from kth to $(k+1)$th step of the outer loop requires potentially v times inner loop iteration. The specific procedure is depicted as follows. When the kth step solutions are achieved $\left(\mathbf{x}^{(k)}, \mathbf{y}^{(k)}, z^{(k)}\right)$, the objective function $f_i(\mathbf{x})$ is substituted by a certain convex function $\tilde{f}_i^{(k,0)}(\mathbf{x})$. The inner loop iteration is only conducted when $\tilde{f}_i^{(k,0)}\left(\hat{\mathbf{x}}^{(k,0)}\right) < f_i\left(\hat{\mathbf{x}}^{(k,0)}\right)$. Otherwise, $\left(\mathbf{x}^{(k+1)}, \mathbf{y}^{(k+1)}, z^{(k+1)}\right) = \left(\hat{\mathbf{x}}^{(k,0)}, \hat{\mathbf{y}}^{(k,0)}, \hat{z}^{(k,0)}\right)$, where $\left(\hat{\mathbf{x}}^{(k,0)}, \hat{\mathbf{y}}^{(k,0)}, \hat{z}^{(k,0)}\right)$ is the optimal solution of the subproblem. The new convex function for the inner loop iteration $\tilde{f}_i^{(k,1)}\left(\mathbf{x}^{(k)}\right)$ which is more conservative compared with $\tilde{f}_i^{(k,0)}\left(\mathbf{x}^{(k)}\right)$, i.e. $\tilde{f}_i^{(k,1)}\left(\mathbf{x}^{(k)}\right) > \tilde{f}_i^{(k,0)}\left(\mathbf{x}^{(k)}\right)$ for all $\mathbf{x} \in X^{(k)}$, except for $\mathbf{x} = \mathbf{x}^k$, $\tilde{f}_i^{(k,1)}\left(\mathbf{x}^{(k)}\right) = \tilde{f}_i^{(k,0)}\left(\mathbf{x}^{(k)}\right)$. Similar to the last inner loop solution, when the optimal solution of the current loop is obtained as $\left(\hat{\mathbf{x}}^{(k,1)}, \hat{\mathbf{y}}^{(k,1)}, \hat{z}^{(k,1)}\right)$, and $\tilde{f}_i^{(k,1)}\left(\hat{\mathbf{x}}^{(k,1)}\right) \geq f_i\left(\hat{\mathbf{x}}^{(k,1)}\right)$ for $i = 0, 1, \ldots, m$, the next out loop iteration point is $\left(\mathbf{x}^{(k+1)}, \mathbf{y}^{(k+1)}, z^{(k+1)}\right) - \left(\hat{\mathbf{x}}^{(k,1)}, \hat{\mathbf{y}}^{(k,1)}, \hat{z}^{(k,1)}\right)$. Otherwise a same inner loop iteration is repeated until $\tilde{f}_i^{(k,v)}\left(\hat{\mathbf{x}}^{(k,v)}\right) \geq f_i\left(\hat{\mathbf{x}}^{(k,v)}\right)$, where $\left(\hat{\mathbf{x}}^{(k,v)}, \hat{\mathbf{y}}^{(k,v)}, \hat{z}^{(k,v)}\right)$ is the inner loop optimal solution of the vth step.

5.4.6 Optimization Framework

Figure 5-11 illustrates the optimization framework. After the prerequisite settings for the finite element analysis, the definition of the objectives and constraints, selection of the design domain, and the initialization can be conducted by applying a unified value for all the design variables. For example, the density of each element in the design domain is 0.8. Then the material property, i.e. BH curve of each element in the design domain can be achieved for the conduction of the finite element analysis. The performance and sensitivity calculations are based on the frozen permeability after the nonlinear solution. The convergence criteria can be set according to the variation ratio of design variables or the objectives.

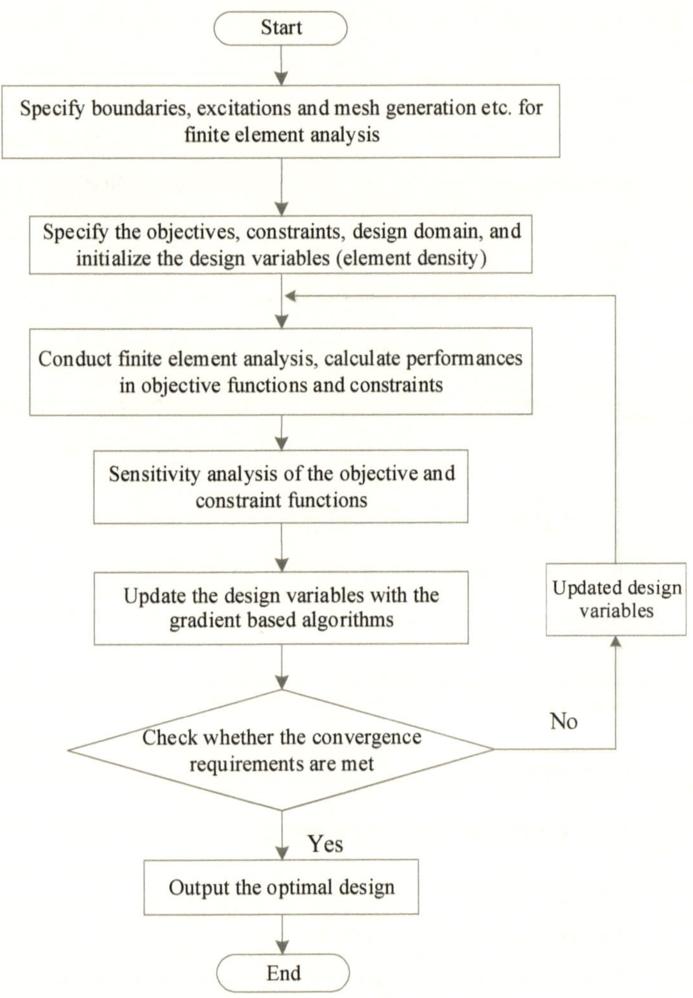

Figure 5-11 Flowchart of the topology optimization.

5.4.7 Postprocessing

Due to the problems of numerical instability and mesh dependency, the gray elements and coarse edges may exist in the optimization result. The gray elements mean that the density of the element is in the interval of (0,1], for example, 0.6. This case is unacceptable since we have to promise the material of the element is either void or solid ferromagnetic material instead of the fictitious material with hypothetical permeability. Various filter techniques are introduced to eliminate the gray elements, such as the field for the sensitivity or the density directly [1]. In this research, for the possible gray elements in the final optimal design, we utilize practical post-processing for achieving the feasible solution. With the interpolation method introduced in section 5.4.1, taking the case when the relative permeability of ferromagnetic material is 3000, and p = 1 as an example, the relative permeability relationship with respect to the element density is demonstrated in Figure 5-12. The relative permeability of an element with a density less than 0.9 is very close to 1 (relative permeability of air). Therefore, in the final post-processing of the optimal solution, the elements with the density lower than 0.9 can be set as zero while others set as 1. The threshold density can be adjusted according to the permeability, penalty factor and the objectives of the solid topology. For the coarse edges of the optimized topology, the Spline can be applied for smoothing the contour to obtain the design suitable for manufacturing.

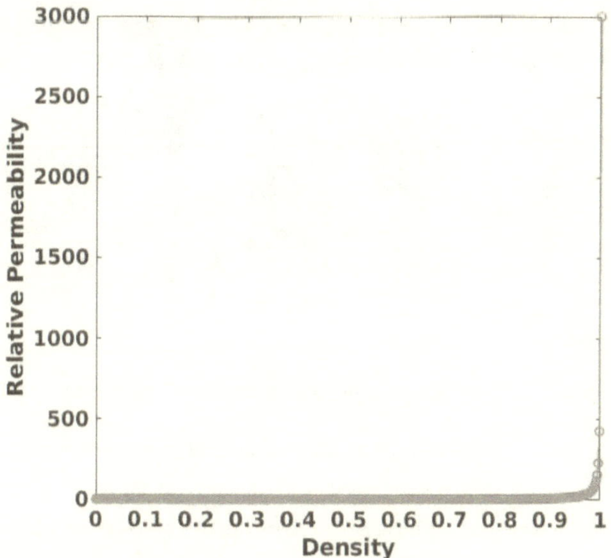

Figure 5-12 Relative permeability v.s. density when $p = 1$.

5.5 Design Example

As presented in the last two sections, the calculation of magnetic field energy, flux linkage, inductance, torque, and their sensitivities are derived. The optimization of the parameters and performance can be conducted by the proposed framework. Moreover, the other related characteristics can also be optimized based on the above sensitivity derivations, such as the cogging torque and back EMF.

To illustrate the effectiveness of the proposed optimization approach, two design examples are presented in this section. The first one is a lightweight rotor design, which is similar to the problem in mechanical engineering. The other is about the torque profile improvement which synthesizes the performance calculation and sensitivity analysis in this research.

5.5.1 Lightweight Rotor Optimization of an SPMSM

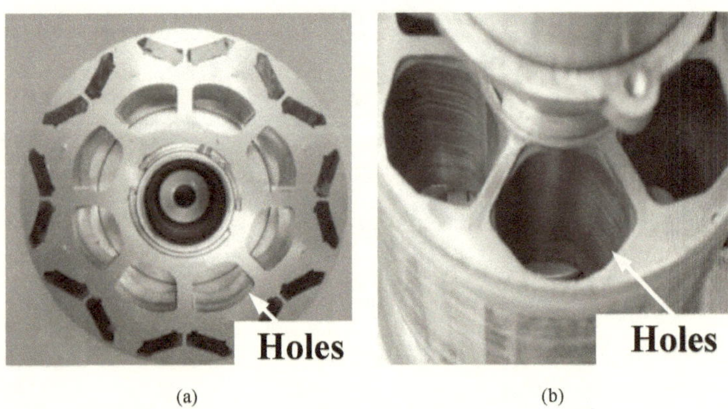

Figure 5-13 Rotor design of (a) Toyota Prius 2010 drive motor, and (b) a servo motor.

The lightweight design optimization of the soft magnetic component of the motors is widely pursued by designers and manufacturers, e.g. the rotor design with large holes of some commercial electrical motors for vehicles such as the Toyota Prius 2010 and servo motors as illustrated in Figure 5-13 [46, 47]. The performance improvement of the design includes but is not limited to the increase in power density. In particular, the dynamic performance, such as the start and brake response speed, can be improved if the mass of the rotor is reduced because of the decreased rotational inertia. Moreover, other performance, e.g. the cooling, may also be enhanced.

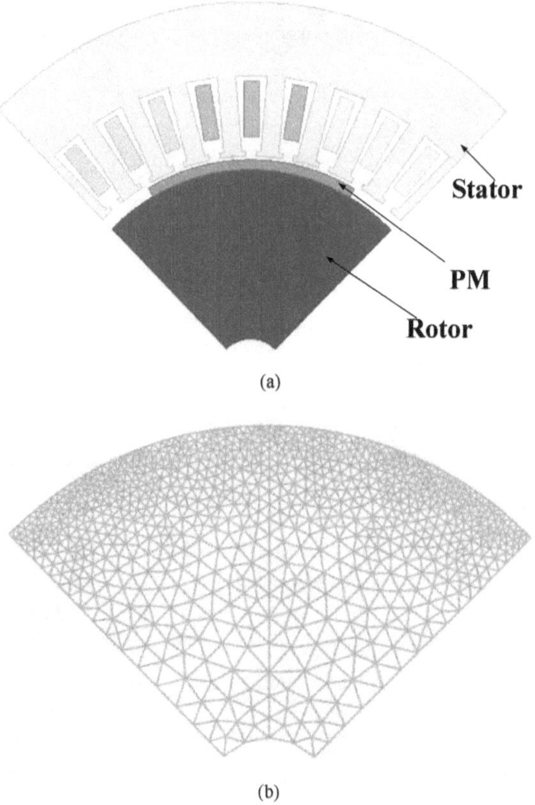

Figure 5-14 (a) 1/4 model of the electrical machine, and (b) elements in the design domain.

The objective function of the design example is simple, which aims to obtain the maximal average torque under the constraint of limited ferromagnetic component volume. If only the electromagnetic performance is considered for the designer with expertise in the electromagnetic analysis, the design optimization problem is simple. However, the aim of the example includes but is not limited to the automatic optimization which can also be conducted with the topology optimization of other disciplinary problems, such as the mechanical and thermal performance simultaneously.

Table 5-1 lists the specifications of the design example. Figure 5-14 illustrates the 1/4 model of the motor and the elements generated in the design domain. The number of elements, i.e. the design variable quantity, is chosen as 1670. For this example, we aim to find the maximum average rated output torque with 60 percent of the rotor core material.

The objective and constraint can be expressed as

$$\min: f = -\frac{1}{N}\sum_{i=1}^{N} T\left(\theta_{\text{elec},i}\right)$$

$$\text{sub.} \quad g = \sum V_e \rho_e - 0.6V \leq 0$$

(5-136)

where V_e is the volume of the element, V the total volume of the design domain and ρ_e is the density of eth element i.e. the eth design variable. The initial values of the design variables are set as 0.6.

Table 5-1 Specification of the design example

Performance	Unit	value
Rated power	kW	9.5
Rated phase current RMS	A	140
Number of poles		4
Rated torque	Nm	59.5
Rated speed	rpm	1500
Advanced current angle	degree	0

The torque calculation takes advantage of the field circuit method by applying the method in the synchronous rotating reference frame. The sensitivities of the objective and constraint can then be obtained as

$$\frac{\partial f}{\partial \rho_e} = -\frac{1}{N}\sum_{i=1}^{N}\frac{\partial T\left(\theta_{\text{elec},i}\right)}{\partial \rho_e}$$

$$= -\frac{1}{N}\sum_{i=1}^{N} 1.5 n_p \left[\frac{\partial \psi_d\left(\theta_{\text{elec},i}\right)}{\partial \rho_e} i_q - \frac{\partial \psi_q\left(\theta_{\text{elec},i}\right)}{\partial \rho_e} i_d\right]$$

(5-137)

$$\frac{\partial g}{\partial \rho_{e}} = V_{e} \qquad (5\text{-}138)$$

The OC method is adopted as the optimization algorithm for this problem with a simple volume constraint.

Figure 5-15 illustrates the change of average output torque and design domain volume with the iteration of the optimization process. Figure 5-16 shows the design configuration at different iteration steps with clear topological changing trends. By the OC method, the volume constraint is well met in each iteration step. Meanwhile, the average output torque reaches the same level with the original design from a low value due to the initial assumption of the density. The optimal result proves that similar output can be achieved with 60% of the rotor material. Figure 5-17 illustrates the phase back EMFs of the initial design, optimized design and smooth design. Since the EMF waves of three phases in one design share the same shape, we just present wave of phase one in initial design, phase two in optimized design, and phase two in smooth design in one figure for comparison. As shown, the reduction of ferromagnetic material in the rotor will not affect the characteristics of the original magnetic circuit significantly.

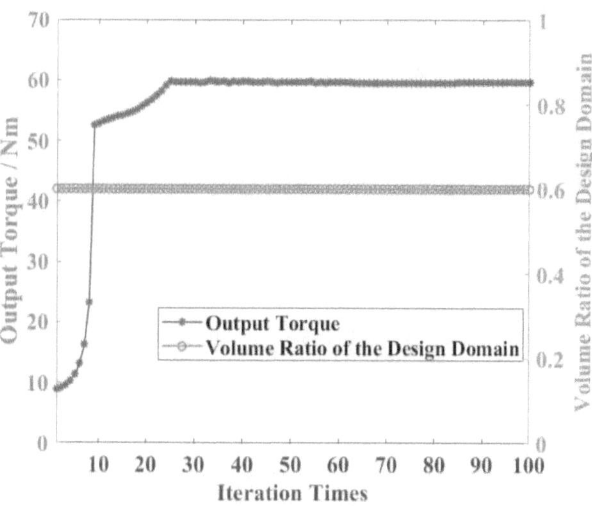

Figure 5-15 Output torque and volume of the design domain iteration history.

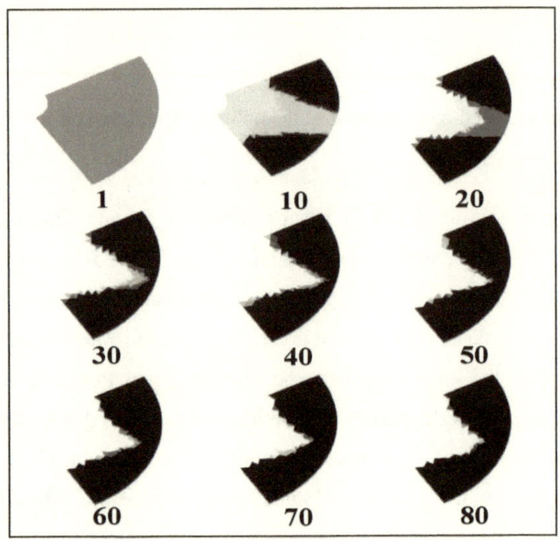

Figure 5-16 Rotor configurations at different iteration steps.

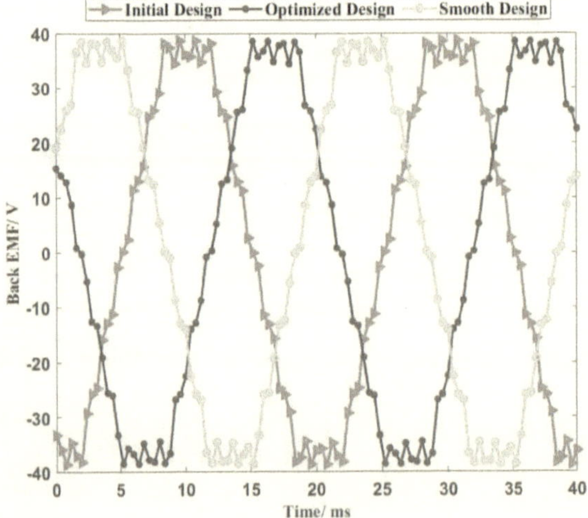

Figure 5-17 Back EMF comparison.

Figure 5-18 shows the flux density maps of the initial design, the optimized raw design, and the smooth design after the postprocessing. According to the topology optimization result, the design domain of low flux density has been deleted to meet the requirement of the constraint for a high-density design.

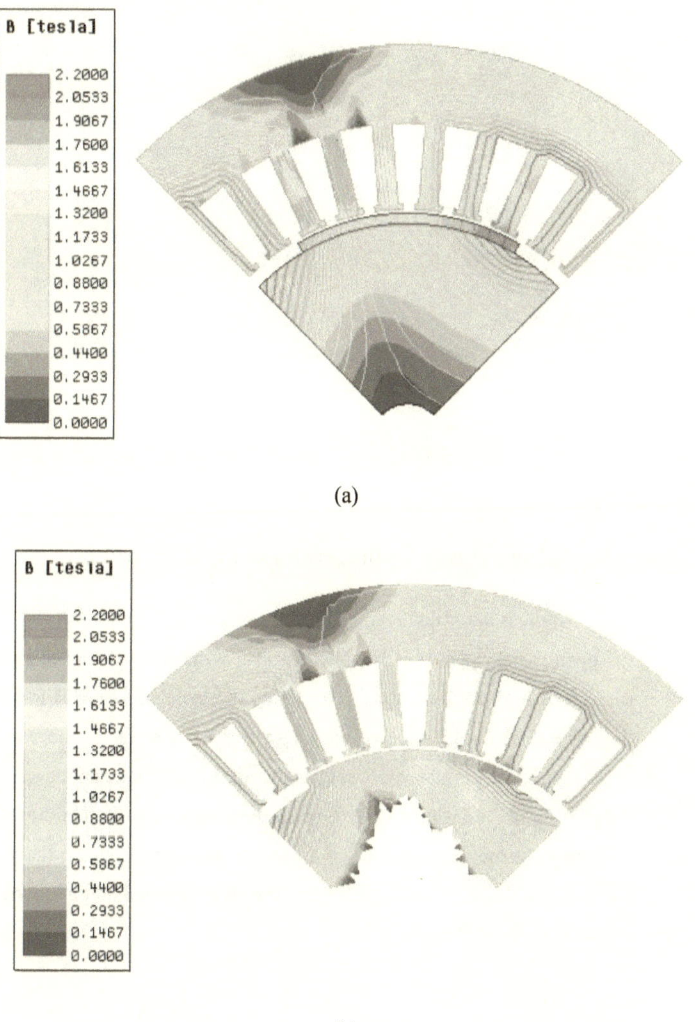

(a)

(b)

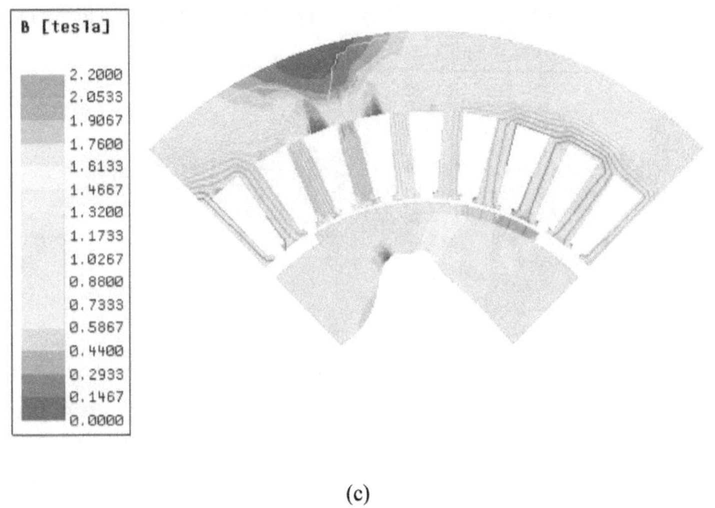

(c)

Figure 5-18 Flux density maps of (a) Initial design, (b) optimized design, and (c)smooth design.

5.5.2 Torque Ripple optimization of an IPMSM

The example is based on the interior permanent magnet motor utilized in the drive system of Toyota Prius 2004 [48]. The specifications of the motor are listed in Table 5-2. In order to achieve high torque output considering the synchronous PM torque and reluctance torque, the advanced angle of the phase current is set as 30 degrees (electrical). The ferromagnetic material component of the rotor as the design domain, while the density of the element in the design domain is set as the design variable. The aim of the optimization is to improve the torque profile of the motor including reducing the torque ripple while not reducing the average torque output when it is operated unidirectionally. The objective and constraint are defined as

$$\min : f = \sum_{h=1}^{M} \left(T_{\mathrm{A,h}}^2 + T_{\mathrm{B,h}}^2 \right)$$
$$sub. \quad g = 260 - \frac{1}{N} \sum_{i=1}^{N} T\left(\theta_{\mathrm{elec},i} \right) \le 0 \tag{5-139}$$

where $T_{A,h}^2$ and $T_{B,h}^2$ are the square of the hth torque harmonics of the torque obtained by DFT with respect to the angular displacement in electrical degrees. The constraint is to limit the average torque no less than 260 Nm.

Table 5-2 Specifications of the motor

Performance	Unit	value
Number of phases		3
Number of poles		8
Number of teeth		48
Rotation speed	rpm	1500
Advanced angle	Electrical Degree	30
Phase current RMS	A	140

According to the performance derivation by the field circuit coupling method, the detailed torque profile is related to the electrical energy and the coupled magnetic energy with respect to the rotational displacement. The torque ripple can be derived from both input and coupling factor fluctuation. Therefore, to achieve a relatively accurate torque calculation for torque ripple minimization, the torque derivation by the field circuit coupling method II is adopted and then the sensitivity of the objective function can be expressed as

$$\frac{\partial f}{\partial \rho_e} = 2\mathbf{T_A}\left(\frac{\partial \mathbf{T_A}}{\partial \rho_e}\right)^{\mathbf{T}} + 2\mathbf{T_B}\left(\frac{\partial \mathbf{T_B}}{\partial \rho_e}\right)^{\mathbf{T}} \tag{5-140}$$

where the harmonic coefficient vectors $\mathbf{T_A}$ and $\mathbf{T_B}$ can be calculated and expressed similarly as (5-104) and (5-105). Take the partial derivation of $\mathbf{T_A}$ as an example, and it can be derived as

$$\frac{\partial \mathbf{T_A}}{\partial \rho_e} = \frac{2}{N}\frac{\partial \mathbf{T}}{\partial \rho_e}\cos\left(\mathbf{\theta^T M}\right) \tag{5-141}$$

where

$$\frac{\partial \mathbf{T}}{\partial \rho_e} = \left[\frac{\partial T\left(\theta_{\text{elec},0}\right)}{\partial \rho_e}, \frac{\partial T\left(\theta_{\text{elec},1}\right)}{\partial \rho_e}, ..., \frac{\partial T\left(\theta_{\text{elec},N-1}\right)}{\partial \rho_e} \right] \tag{5-142}$$

Meanwhile, the sensitivity derivation of the average torque in the last example can still be used in the current constraint. The MMA is applied as the optimization algorithm.

Table 5-3 Torque profile comparison

Torque Performance	Unit	Initial	Optimized	Smooth
Objective	--	439.0	72.4	90.5
Peak to peak value	Nm	50.4	20.3	24.5
Average torque	Nm	265.8	263.5	264.2
Amplitude of 6th harmonic	Nm	8.5	6.0	6.1
Amplitude of 12th harmonic	Nm	19.1	3.0	6.6

The simulation of the machine has been performed on various FEM software packages [49]. It is found that the estimated performance parameters calculated from the FEM-based method are well aligned with the experimental results, such as the back EMF and torque. Therefore, it is reliable using it to optimize the investigated drive system. The parameter setting in this simulation followed the instructions offered in [49], and the optimized topology is verified by ANSYS.

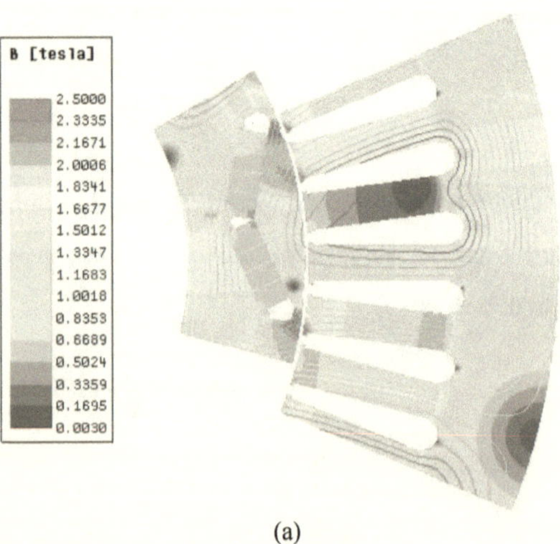

(a)

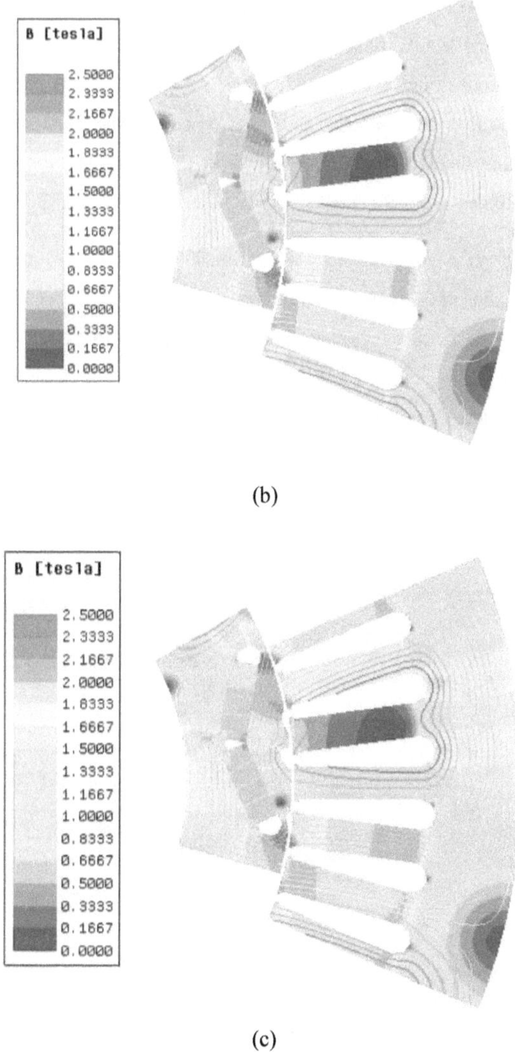

(b)

(c)

Figure 5-19 Flux density maps of (a) Initial design, (b) Optimized design, and smooth design.

Figure 5-19 illustrates the flux density maps of the initial design, optimized design, and smoothed design. Since the initial motor design is of high-power density, with the average torque constraint, nonmagnetic notches appear on the edge of the rotor for reducing the fluctuations of co-energy and input power. Figure 5-20 shows the torque at different angular displacement in electrical degrees and the torque amplitudes of different harmonics after DFT, respectively. Table 5-3 lists the torque profile comparison.

The objective of design decreased from 439 to 72.4 while the peak-to-peak value of torque was reduced from 50.4 to 20.3. This means the fluctuation of output torque dropped from 19.0% to 7.7%. Compared with the mitigation of torque ripple, the change of average torque can be neglected. According to the DFT result, the 6th and 12th harmonics of the output torque are suppressed, which are the main component of ripple.

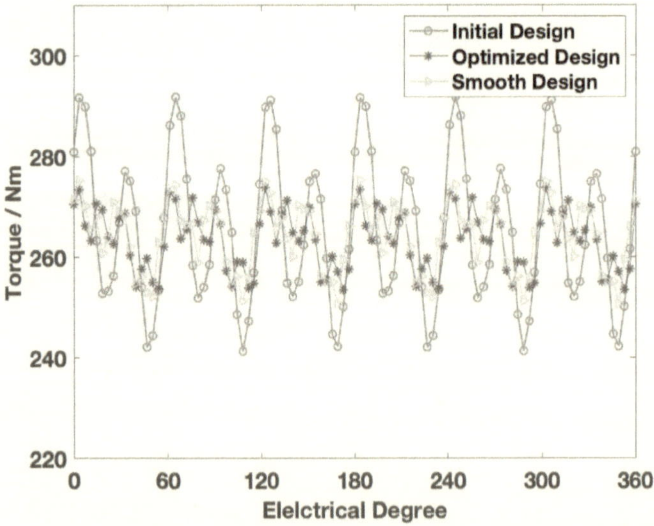

Figure 5-20 Output torque comparison of the initial, optimized, and smooth design.

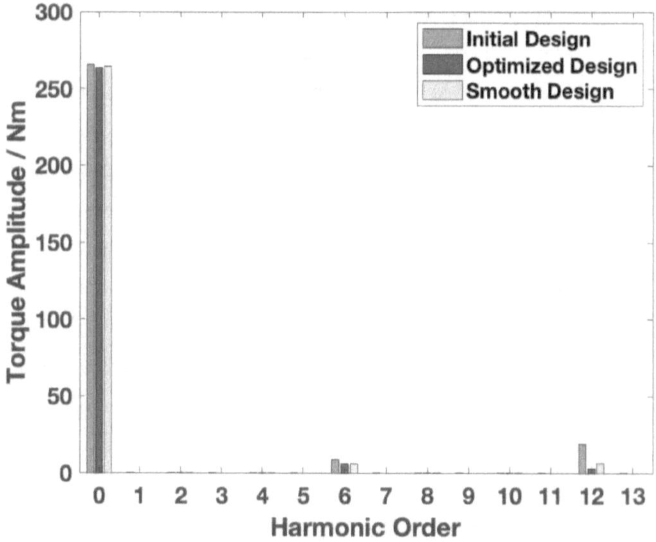

Figure 5-21 Output torque harmonic comparison of the initial, optimized and smooth design.

5.5.3 Comparison and Summary

For the widely applied parametric optimization methods and shape optimization methods by adjusting the nodes on edges, the number of design variables is usually around or less than 10 [50], meaning that the room for performance improvement is limited. The topology optimization takes the material property of each element as the design variable (e.g. air or ferromagnetic material in this paper), so the number of design variables is equal to the number of elements. For example, the number of design variables in the first example is 1670 in the design domain. Therefore, topology optimization has much higher freedom of design optimization. From another point of view, the topology optimization is proposed to relieve the pressure on designers in bringing up new shape designs and optimize them by a more general and automatic process.

From the perspective of computational burden, the computing power cost usually depends on the parameter and performance calculation and iteration of optimization algorithms. When the FEM is applied for parameter and performance evaluation, the parametric and shape optimization with intelligent algorithms usually requires tens or hundreds of times parameter and performance calculation in each iteration. Even though various techniques have been developed for improving the efficiency, such as the design of experiment for reducing the number of samples, surrogate models for replacing the FEM, and design space reduction by sensitivity analysis of the design parameters, hundreds of times of FEM calculation are required in the optimization process. For example, for an 8-parameter optimization problem in [50], 604 FEM calculations are required. In contrast, the topology optimization method in this research utilizes the gradient-based algorithms, which have fast convergence. With the explicit sensitivity formulation, in each iteration, only one FEM calculation is required. Therefore, the number of calculations by FEM is equal to the number of iteration steps. For example, in the first example of this paper, the algorithm convergences at about the 30th iteration, which means that only 30-time calculations by FEM are required. Even though the ability of convergence to the global optimal solution can hardly be promised for the gradient-based algorithms, it can be compensated by setting multiple initial designs while still keeping the comparable computation requirement with the parametric or shape optimization.

Table 5-4 FEM of the examples

Example	No. of elements	No. of nodes	No. of design variables	No. of rotor positions
SPMSM	7436	3924	1670	10
IPMSM	3450	1890	758	50

Table 5-5 Computation time of the optimization

Computation time (s)	Sensitivity	FEM	Variable updating	Each iteration
SPMSM	4.1	66.5	0.31	70.8
IPMSM	316.8	212.7	0.11	532.6

To prove the efficiency of the proposed method, the FEM and computation time of the two design examples are listed in Table 5-4 and Table 5-5 Computation time of the optimization. The optimization is conducted on a laptop with 2.3-GHz Intel Corei5-5300U. Due to the application of the gradient-based optimization algorithms, the computation cost should be considered from the aspects of FEM calculation, sensitivity estimation, and update of the design variables by the algorithm. In contrast, if the derivation and difference method are not applied for the sensitivity calculation, the number of FEM calculations is double that of the design variables. For example, the sensitivity estimation time of example one equals $1670 \times 2 \times 66.5$ s, i.e. 61.7 hours, in one iteration. With the derived sensitivity calculation method in this paper, the sensitivity information computation times of the two examples in one iteration are 4.1 s and 316.8 s, respectively. The computational burden can be effectively reduced and the optimization can be conducted with high efficiency.

5.6 Conclusions

In this research, the topology optimization of the ferromagnetic components of rotors in electrical machines based on the density method is presented. Since the parameters and performance, as well as their sensitivities, are derived based on the energy conversion theory, the presented method is also applicable to other types of electrical machines, and other ferromagnetic components, e.g. those in the stators. Two design examples have been presented to illustrate the effectiveness of the method for lightweight high torque density design and electromagnetic performance enhancement. In the example of torque ripple mitigation, the comprehensive analysis of its derivation from the electromagnetic field is illustrated. It also offers an opportunity for the optimization of single field performance, such as the cogging torque ripple due to the PM field variation with the mechanical movement and the EMF ripple due to the flux linkage variation. The future work will include the topology optimization of multiple components of different materials under different work conditions while considering more parameters and performances.

Chapter 6 Robust Optimization for Electrical Machines Considering Different Uncertainties

6.1 Introduction

As reviewed in Chapter 2, the uncertainty quantification methods can be generally classified as stochastic and non-stochastic methods. For uncertain parameters with sufficient data points, e.g. dimension parameters of a product in mass production, they can be described by a specific stochastic distribution. Under this condition, stochastic methods can be applied to evaluate the robustness by the statistic properties such as the mean and variance of the structural performance. In some circumstances, the uncertain parameters can only be described by their interval bounds since there are not enough samples for achieving the probability distributions. For example, in the early stage of a design process, the designers can hardly obtain the distributions of uncertain parameters, but the bound information such as the dimension tolerances may be available. In this case, the interval methods as one of the non-stochastic approaches can be applied to calculate the upper and lower bounds of performance fluctuation, which reflect the extreme conditions. Then, the worst case is utilized as an indicator of robustness. The conventional uncertainty quantification methods for problems with one type of uncertainties, such as the scan approach (scan the samples within the variation interval to determine the performance variation range) for the interval uncertainty and the Monte Carlo method for the stochastic uncertainty, are frequently used for performance perturbation quantification. Due to the demand for a large number of sampling points, the computation burden is heavy, especially for time-consuming models. For the robust optimization of electrical machines, the calculation cost would increase further due to the required large design population and optimization iterations to apply the intelligent algorithms.

On the other hand, the scenario with both stochastic and interval variables in the robust electrical machine optimization is rarely investigated for the optimization of electrical machines. The above-mentioned approaches applied in the situation of a single type of uncertainty cannot estimate the reliability properly for problems with both stochastic and interval uncertainties. The computational burden is also extremely heavy if the scan and Monte Carlo approach (SMCA) is applied sequentially for the hybrid uncertainty analysis.

Considering cases with stochastic, interval, and their hybrid uncertainties, this research introduces the uncertainty quantification methods into the robust optimization which include PCE and univariate dimension reduction method stochastic uncertainties, and Chebyshev interval (CI) method for interval uncertainties, Polynomial chaos Chebyshev interval (PCCI) method for hybrid uncertainties (with both stochastic and interval uncertainties). All the uncertainty quantification methods mentioned above are integrated into a unified robust optimization framework. To investigate the feasibility of the proposed optimizer, a brushless DC motor design optimization benchmark is investigated. The numerical results prove the effectiveness of the proposed method for robust optimization.

6.2 Problem Definition

In order to present the difference between the deterministic and robust optimization modeling, the deterministic model is rewritten here

$$\begin{aligned}
\min : f &= f(\mathbf{d}) \\
\text{s.t.} \quad g_v(\mathbf{d}) &\leq 0, \\
v &= 1, 2, \cdots, m
\end{aligned} \tag{6-1}$$

where $\mathbf{d}$ is the design variables, $f(\mathbf{d})$ represents the objective function, and $g_v(\mathbf{d})$ represents the vth constraint. In the robust optimization, considering the uncertainties, the problem can be abstracted as

$$\begin{aligned}
\min : f(\mathbf{d}, \zeta) & \\
\text{s.t.} \quad g_v(\mathbf{d}, \zeta) &\leq 0, \\
v &= 1, 2, \cdots, m
\end{aligned} \tag{6-2}$$

where ζ is the uncertainty variable vector. Particularly, it should be noted that the design variables can also be with uncertainty.

In the generalized robust optimization (considering the features RBDO and RDO), the robustness is related to the sensitivity of the performance and whether the constraints are violated with respect to the uncertain variables. Even robustness modeling varies considering different kinds of uncertainties, the discussion of this study tries to utilize the unified general model as presented below.

6.2.1 Robust Optimization Model with Different Uncertainties

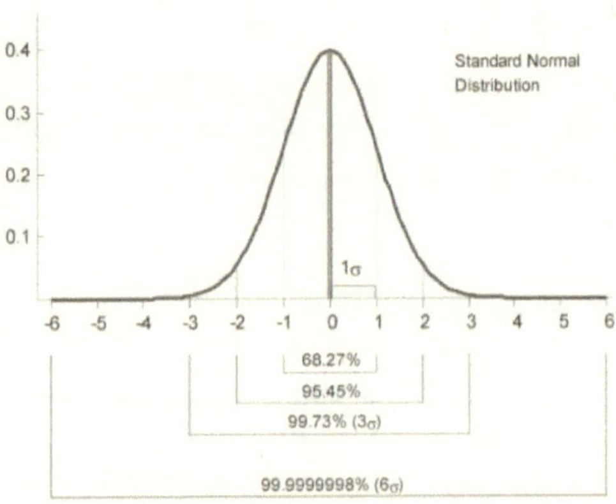

Figure 6-1 Sigma level and its equivalent probability for a normal distribution.

When only stochastic (supposing the uncertainties obey the normal distribution) uncertainties exist, the robust optimization model can be described as

$$\begin{aligned}
&\min : \mu_f(\mathbf{d},\mathbf{X}) + n_f \sigma_f(\mathbf{d},\mathbf{X}) \\
&\text{s.t. } \mu_{g_v}(\mathbf{d},\mathbf{X}) + n_g \sigma_{g_v}(\mathbf{d},\mathbf{X}) \leq 0, \\
&\quad v = 1, 2, \cdots, m \\
&\quad \mathbf{X} \sim N\left(\boldsymbol{\mu}_x, \boldsymbol{\sigma}_x^2\right)
\end{aligned} \tag{6-3}$$

where μ_f, σ_f, μ_{g_k}, and σ_{g_k} are the mean and standard deviations of the objective and constraint, respectively. $\boldsymbol{\mu}_x$ and $\boldsymbol{\sigma}_x$ are the mean and standard deviations of the stochastic uncertainties $\mathbf{X}$, n_f is the weighting factor for the diversity of objectives. Considering the reliability of the constraints, the DFSS approach (a specific case of the moment matching method) is commonly used, i.e. by setting the factor n_g=6 [1, 2]. Figure 6-1 illustrates the sigma level and its equivalent probability for the standard normal distribution. In the case of DFSS optimization, a large probability of the product in mass production can be promised without violating the constraints.

When there exist only the interval uncertainties, the robust optimization is modeled by the worst case, i.e. the upper bounds of the objective function and constraints expressed as

$$\begin{aligned}
&\min : \max f(\mathbf{d}, \mathbf{Y}) \\
&\text{s.t.} \quad \max g_v(\mathbf{d}, \mathbf{Y}) \le 0, \\
&\qquad v = 1, 2, \cdots, m \\
&\qquad \lambda - \delta \le \mathbf{Y} \le \lambda + \delta
\end{aligned} \tag{6-4}$$

where $\mathbf{Y}$ is the interval uncertainty vector with nominal λ and length δ from bounds to the nominal.

In the case of the hybrid uncertainties, the objective functions and constraints have the characteristics of both the stochastic and interval uncertainties. The robust optimization considering the hybrid uncertainties can then be expressed as

$$\begin{aligned}
&\min : \max \left[\mu_f(\mathbf{d}, \mathbf{X}, \mathbf{Y}) + n_f \sigma_f(\mathbf{d}, \mathbf{X}, \mathbf{Y}) \right] \\
&\text{s.t.} \quad \max \left[\mu_{g_k}(\mathbf{d}, \mathbf{X}, \mathbf{Y}) + n_g \sigma_{g_k}(\mathbf{d}, \mathbf{X}, \mathbf{Y}) \right] \le 0, \\
&\qquad k = 1, 2, \cdots, q \\
&\qquad \mathbf{X} \sim N\left(\mathbf{\mu}_x, \mathbf{\sigma}_x^2 \right) \\
&\qquad \lambda - \delta \le \mathbf{Y} \le \lambda + \delta
\end{aligned} \tag{6-5}$$

To estimate the extreme values of objectives and constraints, the mean and standard deviation are calculated separately by SMCA as illustrated in Figure 6-2, where μ_{ub}, σ_{ub}, μ_{lb}, σ_{lb} are the calculation of the upper bound and lower bound mean and standard deviation. Then, the worst-case under this situation can be estimated as the sum of the upper bounds of mean and standard deviation multiplied by the sigma level factor and (6-5) can be rewritten as

$$\min : \mu_{ub}\left[f(\mathbf{d},\mathbf{X},\mathbf{Y})\right]+n_f\sigma_{ub}\left[f(\mathbf{d},\mathbf{X},\mathbf{Y})\right]$$

$$\text{s.t.} \quad \mu_{ub}\left[g_k\,\mathbf{d},\mathbf{X},\mathbf{Y}\right]+n_g\sigma_{ub}\left[g_k\left(\mathbf{d},\mathbf{X},\mathbf{Y}\right)\right]\le 0,$$

$$k=1,2,\cdots,q \tag{6-6}$$

$$\mathbf{X}\sim N\left(\boldsymbol{\mu}_x,\boldsymbol{\sigma}_x^{\,2}\right)$$

$$\boldsymbol{\lambda}-\boldsymbol{\delta}\le \mathbf{Y}\le\boldsymbol{\lambda}+\boldsymbol{\delta}$$

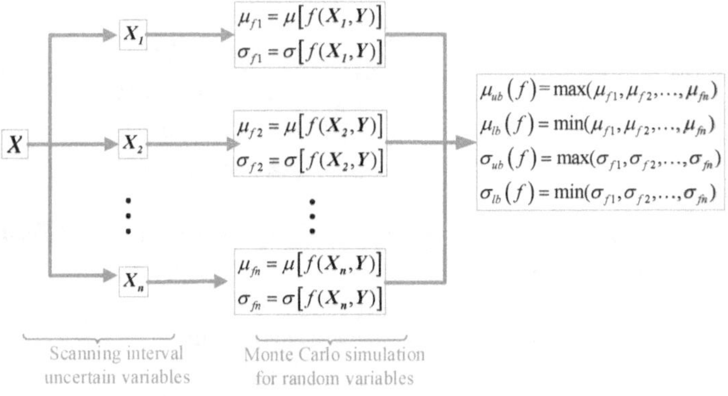

Figure 6-2 Estimation process of extreme values of mean and standard deviation for a problem with hybrid uncertainty.

6.2.2 Uniformity of the Robust Optimization Models

Even the robust model varies regarding different kinds of uncertainties, we can see the uniformity of the models which can be recognized as the generalized worst-case model. The modeling with stochastic uncertainties utilizes the performance $\mu_{g_k}+6\sigma_{g_k}$ as the worst case for promising the design meeting the constraints of a large probability. The same is also with the hybrid robust optimization model.

Regarding the sensitivity of the objective function with respect to the uncertainties, it can be represented and quantified by the standard deviation, variation interval for the problems with stochastic and interval uncertainties, respectively. For the hybrid case, the objective variation interval can be calculated as $\mu_{ub}(f)-\mu_{lb}(f)+2n_f\sigma_{ub}(f)$. The robustness can be considered in a single objective form as (6-3) with the weighting factor. It can also be processed individually in the objective (as a multi-objective problem) or limited in the constraints. Even though it is not mainly discussed in this study, it should be noticed that the methods introduced below for the uncertainty quantification have the ability on the effective objective robustness calculation. In the design example, there are comparisons between the methods we introduced here with the classical methods on the calculation of the factor, such as the standard deviation comparison calculated by MCA and PCE.

6.3 Methods for Stochastic Uncertainties

6.3.1 Polynomial Chaos Expansion

The PCE [3] method is to express a stochastic process with the sum of specific orthonormal polynomials of the independent variables. With the established polynomial expansion, the statistical moments, i.e. the mean and variance of the responses can be easily achieved. Compared with the MCA, the establishment of PCE with limited orders, i.e. calculate the coefficients of the polynomial, usually requires fewer samples. Consequently, the sampling calculation time can be effectively saved. Taking the Gaussian process as an example, the Hermite polynomial expansion is proposed by Wiener, which can be expressed as

$$f(\mathbf{X})=c_0+\sum_{i_1=1}^{n}c_{i_1}He_1\left(x_{i_1}\right)+\sum_{i_1=1}^{n}\sum_{i_2=1}^{i_1}c_{i_1i_2}He_2\left(x_{i_1},x_{i_2}\right)+\cdots$$
$$\cdots+\sum_{i_1=1}^{n}\sum_{i_2=1}^{i_1}\sum_{i_3=1}^{i_2}c_{i_1i_2i_3}He_3\left(x_{i_1},x_{i_2},x_{i_3}\right)+\cdots \tag{6-7}$$

where the Hermite polynomial is calculated as

$$He_n\left(x_{i_1},\cdots,x_{i_n}\right)=e^{1/2\mathbf{x}^T\mathbf{x}}(-1)^n\frac{\partial^n}{\partial x_{i_1}\cdots x_{i_n}}e^{-1/2\mathbf{x}^T\mathbf{x}} \tag{6-8}$$

For example, one-dimensional Hermite polynomial can be expressed as

$$He_0 = 1,$$
$$He_1(x) = x,$$
$$He_2(x) = x^2 - 1,$$
$$He_3(x) = x^3 - 3x,$$
$$\vdots$$
$$He_{i+1}(x) = xHe_i(x) - iHe_{i-1}(x)$$

Considering the orthogonality of the Hermite polynomial with variables of standard normal distributions, the truncated expansions of the first orders can be simplified as

$$F_1 = c_0 + \sum_{i=1}^{n} c_i x_i \tag{6-9}$$

$$F_2 = c_0 + \sum_{i=1}^{n} c_i x_i + \sum_{i=1}^{n} c_{i,i}\left(x_i^2 - 1\right) + \sum_{i=1}^{n-1}\sum_{j>i}^{n} c_{i,j} x_i x_j \tag{6-10}$$

$$F_3 = c_0 + \sum_{i=1}^{n} c_i x_i + \sum_{i=1}^{n} c_{i,i}\left(x_i^2 - 1\right) + \sum_{i=1}^{n} c_{i,i,i}\left(x_i^3 - 3x_i\right) + \cdots$$
$$+ \sum_{i=1}^{n}\sum_{j>i}^{n} c_{i,j} x_i x_j + \sum_{i=1}^{n}\sum_{j=i}^{n} c_{i,j,j}\left(x_i x_j^2 - x_i\right) + \sum_{i=1}^{n-2}\sum_{j>i}^{n-1}\sum_{k>j}^{n} c_{i,j,k} x_i x_j x_k \tag{6-11}$$

The number of the coefficients are

$$N = \frac{(n+p)!}{n!\,p!} \tag{6-12}$$

These polynomials are orthogonal with respect to the weight function

$$\langle He_m(x), He_n(x)\rangle = \int He_m(x)He_n(x)w(x)dx = \delta_{mn}$$
$$w(x) = \frac{1}{\sqrt{(2\pi)^l}} e^{-1/2x^2} \tag{6-13}$$

$$\delta_{mn} = \begin{cases} 1 & (m = n) \\ 0 & (m \neq n) \end{cases} \tag{6-14}$$

The multivariate polynomials are then assembled as the tensor product of their univariate counterparts. The multivariate polynomials constructed are orthonormal:

$$\langle \mathbf{He_m}, \mathbf{He_n} \rangle = \delta_{mn} \tag{6-15}$$

where δ_{mn} is an extension Kronecker symbol to the multi-dimensional case.

Various methods can be applied for calculating the coefficients in the PCE such as the least square method (LSM) and quadrature methods. By using the LSM, the coefficients matrix can be expressed as

$$\mathbf{c} = \left(\mathbf{A}^T \mathbf{A}\right)^{-1} \mathbf{A}^T \mathbf{F(X)} \tag{6-16}$$

where $\mathbf{A}$ is the matrix that contains the values of all the basis polynomials in the experimental design points.

For an objective function $f(\mathbf{X})$ with n probability variables, the kth origin moment $E(f^k)$ can be expressed as:

$$
\begin{aligned}
E(f^k) &= E\left(\left[f(\mathbf{X})\right]^k\right) \\
&= \int_{-\infty}^{\infty}\int_{-\infty}^{\infty}...\int_{-\infty}^{\infty} f^k(\mathbf{X}) p(X_1, X_2, ..., X_n) dX_1 dX_2 ... dX_n \\
&= \int_{-\infty}^{\infty}\int_{-\infty}^{\infty}...\int_{-\infty}^{\infty} f^k(\mathbf{X}) p(X_1) p(X_2)...p(X_n) dX_1 dX_2 ... dX_n
\end{aligned} \tag{6-17}
$$

where $p(X_1, X_2, ..., X_n)$ is the joint probability distribution function of probability uncertain variables. Due to the orthonormality of the polynomial basis, the first two moments of a PCE are encoded in its coefficients

$$\mu(f(X)) = E[f(X)] = c_0 \tag{6-18}$$

$$\sigma[f(X)]^2 = E\left\{[f - \mu(f)]^2\right\} = \sum_{i=1}^{N-1} c_i^2 \tag{6-19}$$

The PCE method has also developed for other kinds of distributions by applying different families of orthonormal polynomials. Table 6-1 lists the categories of the distribution and their orthogonal polynomials.

Table 6-1 Polynomials for different distributions

Distributions	Polynomials
Gaussian	Hermite
Gamma	Laguerre-chaos
Beta	Jacobi-chaos
Uniform	Legendre
Non-negative Binomial	Meixner
Binomial	Krawtchouk
Hypergeometric	Hahn
Poisson	Charlier

6.3.2 Dimension Reduction Method

Compared with the PCE method, the dimension reduction method offers a different solution way by transferring the multivariate model into a series of low univariate problems.

Using the univariate dimension reduction method, the multidimensional problem will be transformed into a series of one-dimension problems. As a result, the multidimensional Gauss integral turns to a one-dimension Gauss integral. Then the compliance can be approximated as follows:

$$f(\mathbf{X}) \cong \sum_{j=1}^{n} \hat{f}_j(\mathbf{X}) - (n-1) f(u_1, u_2, \cdots, u_{\mathrm{n}}) \tag{6-20}$$

where $\hat{f}_j(\mathbf{X}) = f(u_1, \cdots, u_{j\text{-}1}, X_j, u_{j+1}, \cdots, u_{\mathrm{n}})$, u_j is the mean value of X_j. Then the kth origin moment of compliance is:

$$E(f^k) = E\left(\left[\sum_{j=1}^{n} \hat{f}_j(\mathbf{X}) - (n-1) f(u_1, u_2, \cdots, u_{\mathrm{n}})\right]^k\right) \tag{6-21}$$

Based on the binomial theorem, we can obtain:

$$E(f^k) = \sum_{i=0}^{k} \binom{k}{i} E\left(\sum_{i=1}^{n} \hat{f}_j(\mathbf{X}) \right)^i \left[-(n-1) f\left(u_1, u_2, \cdots, u_n\right) \right]^{k-i} \qquad (6\text{-}22)$$

Define

$$S_j^i = E\left[\left\{ \sum_{i=1}^{j} \hat{f}_j(\mathbf{X}) \right\}^i \right] \qquad (6\text{-}23)$$

in which $j = 1, \cdots, n;\ \ i = 1, \cdots, k\ \ S_j^i$ can be calculated by the following recursion (6-24).

$$S_1^i = E\left[\left(\hat{f}_1(\mathbf{X}) \right)^i \right]; \qquad\qquad i = 1, \cdots, k$$

$$S_2^i = \sum_{l=0}^{i} \binom{i}{k} S_1^l E\left[\left(\hat{f}_2(\mathbf{X}) \right)^{i-l} \right]; \qquad\qquad i = 1, \cdots, k$$

$$\vdots$$

$$S_j^i = \sum_{l=0}^{i} \binom{i}{l} S_{j-1}^l E\left[\left(\hat{f}_j(\mathbf{X}) \right)^{i-l} \right]; \qquad\qquad i = 1, \cdots, k$$

$$\vdots$$

$$S_n^i = \sum_{l=0}^{i} \binom{i}{l} S_{n-1}^l E\left[\left(\hat{f}_n(\mathbf{X}) \right)^{i-l} \right]; \qquad\qquad i = 1, \cdots, k$$

$$(6\text{-}24)$$

$$E(f^k) = \sum_{i=0}^{k} \binom{k}{i} S_n^i \left[-(n-1) f\left(u_1, u_2, \cdots, u_n\right) \right]^{k-i} \qquad (6\text{-}25)$$

According to the Gauss–Hermite quadrature rule, n sampling points will produce the exact integral when $f(\mathbf{X})$ is a polynomial of degree $2n-1$ or less. In the case of the polynomial with a single variable, 3 samples for each variable can usually meet the precision requirement of the modeling. Then the origin moments of the system, or the mean and standard deviation of the objective or constraint functions can be calculated with $3n+1$ samples in total from the origin moments of subsystems with a single variable.

6.4 Method for Interval Uncertainties

Similar to the PCE, the CI method approximate the original interval function $f(\mathbf{Y})$ with interval variables $\mathbf{Y}$ by a truncated Chebyshev series of degree k. The Chebyshev polynomial is one of the orthogonal polynomials constituted with multiplicative cosine functions. It is known the cosine function fluctuates within the interval [-1,1]. Therefore, the Chebyshev polynomials with coefficient 1 have the supremum and infimum, i.e. 1 and -1. The interval arithmetic can then be employed to conveniently compute the interval bounds of the approximated function [4, 5].

Taking the univariate problem as an example, the kth order Chebyshev series is defined as

$$\phi_k(y) = \cos k\theta \tag{6-26}$$

where $y \in [-1,1]$, $\theta = \arccos(y) \in [0, \pi]$.

Generally, for an uncertain variable in the interval $\zeta \in [a,b]$, a transformation can be conducted for obtaining the standard interval variable

$$y = \frac{2\zeta - (b+a)}{b-a} \in [-1,1] \tag{6-27}$$

According to (6-26), the univariate Chebyshev polynomial can be calculated as

$$\begin{aligned}
\phi_0(y) &= 1 \\
\phi_1(y) &= y \\
\phi_k(y) &= 2y\phi_{k-1}(y) - \phi_{k-2}(y), k \geq 2
\end{aligned} \tag{6-28}$$

Similar to Hermite polynomial, Chebyshev polynomial also have orthogonality which can be expressed as

$$\int_{-1}^{1} \phi_k(y)\phi_p(y)w(y)dy = \int_0^{\pi} \cos k\theta \cos p\theta d\theta = \begin{cases} \pi, & k = p = 0 \\ \pi/2, & k = p \neq 0 \\ 0, & k \neq p \end{cases} \tag{6-29}$$

where the weight is

$$w(y) = \frac{1}{\sqrt{1-y^2}} \tag{6-30}$$

The response of a $y \in [-1, 1]$ can be estimated as truncated series with kth order Chebyshev polynomial

$$f(y) \approx \frac{1}{2}c_0 + \sum_{i=1}^{k} c_i \phi_i(y)$$

(6-31)

For the multivariate case, the truncated Chebyshev series with kth order can be written as

$$f(\mathbf{Y}) \approx \sum_{\substack{0 \leq n_1, n_2, \cdots, n_m \leq k \\ n_1 + n_2 + \cdots + n_m \leq k}} c_{n_1 n_2 \cdots n_m} \Psi_{n_1 n_2 \cdots n_m}(\mathbf{Y})$$

(6-32)

where $\mathbf{Y} = [Y_1, Y_2, \ldots, Y_m]$ denotes the interval vector, $c_{n_1 n_2 \cdots n_m}$ represents the corresponding Chebyshev coefficient, and $n_i, i = 1, 2, \ldots, m$ constitutes an index vector.

The multivariate Chebyshev polynomial is the tensor product of the univariate Chebyshev series

$$\Psi_{n_1 n_2 \cdots n_m}(\mathbf{Y}) = \prod_{i=1}^{m} \cos(n_i[\theta_i])$$

$$[\theta_i] = \arccos(Y_i) \in [0, \pi], i = 1, 2, \cdots, m$$

(6-33)

Similar to PCE, LSM can be applied for the coefficient calculation. In order to ensure the numerical stability, the number of the sampling points must be higher than the number of coefficients and the design size is recommended to be set as twice the number of the coefficient for obtaining robust estimates. After the coefficients are achieved, the Chebyshev polynomial model is built. Obviously, the first component of the model is a constant term which is regarded as the nominal value of the model, while the other terms contain interval variables that are fluctuation response of the uncertainties. Based on the characteristics of the cosine functions, we can find

$$\prod_{i=1}^{m} \cos(n_i[\theta_i]) = [-1, 1]$$

(6-34)

This property provides the feasibility for the Chebyshev model in the interval uncertain analysis with high efficiency. The maximum and minimum values (or the bound values) of $f(\mathbf{Y})$ with respect to the interval uncertain variables can be estimated as

$$\max f(\mathbf{Y}) = c_{00...0} + \sum_{\substack{0 \le n_1, n_2, \cdots, n_m \le k \\ n_1 + n_2 + \cdots + n_m \le k, n_1 + n_2 + \cdots + n_m \ne 0}} \left| c_{n_1 n_2 \cdots n_m} \right|$$

$$= c_{00...0} + \sum_{\substack{0 \le n_1, n_2, \cdots, n_m \le k \\ n_1 + n_2 + \cdots + n_m \le k, n_1 + n_2 + \cdots + n_m \ne 0}} c_{n_1 n_2 \cdots n_m} \, \text{sign}\left(c_{n_1 n_2 \cdots n_m} \right)$$

(6-35)

$$\min f(\mathbf{Y}) = c_{00...0} + \sum_{\substack{0 \le n_1, n_2, \cdots, n_m \le k \\ n_1 + n_2 + \cdots + n_m \le k, n_1 + n_2 + \cdots + n_m \ne 0}} -\left| c_{n_1 n_2 \cdots n_m} \right|$$

$$= c_{00...0} + \sum_{\substack{0 \le n_1, n_2, \cdots, n_m \le k \\ n_1 + n_2 + \cdots + n_m \le k, n_1 + n_2 + \cdots + n_m \ne 0}} -c_{n_1 n_2 \cdots n_m} \, \text{sign}\left(c_{n_1 n_2 \cdots n_m} \right)$$

(6-36)

where the sign function is defined as:

$$\text{sign(x)} = \begin{cases} -1, & x < 0 \\ 0, & x = 0 \\ 1, & x > 0 \end{cases}$$

(6-37)

As the method presented above, based on the CI method, the maximum values of the interval objective function $f(\mathbf{d}, \mathbf{Y})$ and the interval constraint functions $g_v(\mathbf{d}, \mathbf{Y})$ with the constant design parameter $\mathbf{d}$ and interval parameter $\mathbf{Y}$ can be estimated by (6-36).

In summary, the CI model is employed to estimate the worst value of the uncertain function. As shown in (6-32), the uncertain function with interval uncertainties is approximated by multivariate Chebyshev polynomials and corresponding coefficients. Based on the characteristics of the cosine functions, the values of the multivariate Chebyshev polynomials vary in [-1,1]. Hence, the extreme value of the uncertain function can be efficiently evaluated by (6-36), where the multivariate polynomials are all considered to be extreme values 1 or -1 and have the same sign with the corresponding coefficients. The solution is relatively conservative than the accurate robust solution, because the multivariate Chebyshev polynomials fluctuate in the interval [-1,1], and can hardly all achieve the extreme value -1 or 1 at the same time due to the uncertain variables' perturbation.

6.5 Method for Hybrid Uncertainties

This section introduces the PCCI method [6, 7] in which the stochastic uncertainties are accounted for by the PCE and the interval uncertainties are quantified by the Chebyshev inclusion functions.

The stochastic and interval uncertainty vectors are noted as $\mathbf{X}$ and $\mathbf{Y}$ in a performance function $P(\mathbf{X}, \mathbf{Y})$. If the uncertainty variables are non-standard, they should be transformed into the standard normal variables firstly, ξ_i $(i = 1, 2, ..., n)$, that belongs to $N(0,1)$ and the standard interval variables, η_j $(j = 1, 2, ..., m)$, that belong to [-1,1]. After transforming the uncertain variables are transformed into the standard form, the performance function is established by the following steps.

As the first step, by considering only the stochastic uncertainty parameters, the performance function $P(\xi, \eta)$ can be expanded by the pth order polynomial as the following

$$P(\xi, \eta) = \sum_{\chi \in \mathbb{N}^n, \|\chi\| \le p} c_\chi(\eta) H_\chi(\xi) \tag{6-38}$$

where χ is the index vector $(\chi_1, \chi_2, \cdots, \chi_n)$ while

$$\begin{aligned}
\|\chi\| &= \chi_1 + \chi_2 + \cdots + \chi_n, \\
\chi_i &= 0, 1, \cdots, p \\
i &= 1, 2, \cdots, n
\end{aligned} \tag{6-39}$$

According to this condition, the term number, i.e. the coefficient number can be obtained as

$$k = C_{n+p}^p \tag{6-40}$$

$H_\chi(\xi)$ and $c_\chi(\eta)$ $\left(\chi \in N^n \text{ and } \|\chi\| \le p\right)$ are the multivariate Hermite polynomial and coefficients, respectively, and $c_\chi(\eta)$ can be obtained by the LSM as

$$\mathbf{c}_\chi(\eta) = \left(\mathbf{A}^\mathrm{T}\mathbf{A}\right)^{-1}\mathbf{A}^\mathrm{T}\mathbf{P}(\xi,\eta) \tag{6-41}$$

$$\mathbf{A} = \begin{bmatrix} H_{\chi^{(1)}}\left(\xi^{(1)}\right) & H_{\chi^{(2)}}\left(\xi^{(1)}\right) & \cdots & H_{\chi^{(k)}}\left(\xi^{(1)}\right) \\ H_{\chi^{(1)}}\left(\xi^{(2)}\right) & H_{\chi^{(2)}}\left(\xi^{(2)}\right) & \cdots & H_{\chi^{(k)}}\left(\xi^{(2)}\right) \\ \vdots & \vdots & \ddots & \vdots \\ H_{\chi^{(1)}}\left(\xi^{(l)}\right) & H_{\chi^{(2)}}\left(\xi^{(l)}\right) & \cdots & H_{\chi^{(k)}}\left(\xi^{(l)}\right) \end{bmatrix} \tag{6-42}$$

where l denotes the number of samples for the stochastic variables.

For the PCE (6-38), the intervals are regarded as fixed values. Then in the second step, the interval uncertainties are considered. By the pth order Chebyshev inclusion function, the interval coefficient $c_\chi(\eta)$ can be approximated as

$$c_\chi(\eta) = \sum_{\gamma \in \mathbb{N}^m, \|\gamma\| \le p} c_{\chi.\gamma} \Psi_\gamma(\eta) \tag{6-43}$$

where γ is a vector of multiple indices consisting of

$$\begin{aligned} \gamma_j &= 0,1,\cdots,p \\ j &= 1,2,\cdots,m \end{aligned} \tag{6-44}$$

indicating the components of the multivariate Chebyshev polynomial, $c_{\chi.\gamma}$ are the coefficients, and

$$\Psi_\gamma(\eta) = \prod_{j=1}^{m} \varphi_{\gamma_j}(\eta_j) \tag{6-45}$$

is the multivariate Chebyshev polynomial built from the univariate polynomial

$$\varphi_{\gamma_j}(\eta_j) = \cos(\gamma_j[\theta_j]) \tag{6-46}$$

$$[\theta_j] = \arccos(\eta_j) = [0,\pi] \tag{6-47}$$

Similar to (6-41) and (6-42), the coefficients $c_{\chi,\gamma}$ can be obtained by the LSM.

From the above, the performance function with hybrid stochastic and interval uncertainties can be expressed by the coefficients, which are functions of the interval uncertainties, and the multivariate polynomials, which are functions of the stochastic variables. Through the polynomial chaos expansion and the Chebyshev inclusion function, the performance function with hybrid uncertainties can be approximated by

$$P(\xi,\eta) = \sum_{\chi \in \mathbb{N}^n,\, \|\chi\| \leq p} \sum_{\gamma \in \mathbb{N}^m,\, \|\gamma\| \leq p} c_{\chi,\gamma} \Psi_\gamma(\eta) H_\chi(\xi) \tag{6-48}$$

in which the PCCI expansion coefficients $c_{\chi,\gamma}$ can be calculated by using the LSM twice as

$$\mathbf{c}_{\chi,\gamma} = \left(\mathbf{B}^T\mathbf{B}\right)^{-1}\mathbf{B}^T\mathbf{P}^T\mathbf{A}\left(\mathbf{A}\mathbf{A}^T\right)^{-1} \tag{6-49}$$

where A and B are the sample matrices with components $A_{rs} = H_{\chi^{(s)}}\left(\xi^{(r)}\right)$, $B_{rs} = \Psi_{\gamma^{(s)}}\left(\eta^{(r)}\right)$, and $\xi^{(r)}$ and $\eta^{(r)}$ are the collocation points of stochastic and interval variables selected from the zeros of the one dimensional higher Hermite polynomials and Chebyshev polynomials, respectively. $P(\xi,\eta)$ is the output vector at the collocation points. For a problem with n stochastic variables and m interval variables, the total polynomial terms will be $C_{n+p}^p C_{m+p}^p$, i.e. the number of the expansion coefficients needs to be calculated, which is also the minimum sampling number. There is also a practical suggestion on the sampling number for improving the coefficient calculation accuracy by increasing the sample number appropriately such as twice the term number.

Finally, based on the orthonormal properties of the polynomials, the mean and interval variance of the performance function can be obtained by

$$\mu(P(\xi,\eta)) = \sum_{\gamma \in \mathbb{N}^m,\, \|\gamma\| \leq p} c_{0,\gamma} \Psi_\gamma(\eta) \tag{6-50}$$

$$\sigma^2(P(\xi,\eta)) = \sum_{\chi \in \mathbb{N}^n,\, \chi \neq 0,\, \|\chi\| \leq p} \left(\sum_{\gamma \in \mathbb{N}^m,\, \|\gamma\| \leq p} c_{\chi,\gamma} \Psi_\gamma(\eta) \right)^2 \tag{6-51}$$

The mean and variance in (6-50) and (6-51) are functions of the interval uncertainties, and they have intervals with lower and upper bounds. Based on the characteristics of the trigonometric functions, we know that $\Psi_0(\boldsymbol{\eta}) = 0$ and $\Psi_{\gamma \in \mathbb{N}^m, \gamma \neq 0}(\boldsymbol{\eta}) = [-1, 1]$. The bounds of the mean and standard variance can thus be easily approximated, as the following

$$\mu(f(\boldsymbol{\xi}, \boldsymbol{\eta})) = \begin{bmatrix} c_{0,0} - \left(\sum_{\gamma \in \mathbb{N}^m, \gamma \neq 0 0, \|\gamma\| \leq p} |c_{0,\gamma}| \right), \\ c_{0,0} + \left(\sum_{\gamma \in \mathbb{N}^m, \gamma \neq 0, \|\gamma\| \leq p} |c_{0,\gamma}| \right) \end{bmatrix} \tag{6-52}$$

$$\sigma(P(\boldsymbol{\xi}, \boldsymbol{\eta})) = \begin{bmatrix} \sqrt{\sum_{\substack{\chi \in \mathbb{N}^n, \chi \neq 0 \\ \|\chi\| \leq p}} \sum_{\substack{\gamma \in \mathbb{N}^m \\ \|\gamma\| \leq p}} c_{\chi,\gamma}^2 - 2 \left| \sum_{\substack{\chi \in \mathbb{N}^n, \chi \neq 0 \\ \|\chi\| \leq p}} \sum_{\substack{\gamma \in \mathbb{N}^m \\ \|\gamma\| \leq p}} c_{\chi,\gamma^{(k_1)}} c_{\chi,\gamma^{(k_2)}} \right|}, \\ \sqrt{\sum_{\substack{\chi \in \mathbb{N}^n, \chi \neq 0 \\ \|\chi\| \leq p}} \sum_{\substack{\gamma \in \mathbb{N}^m \\ \|\gamma\| \leq p}} c_{\chi,\gamma}^2 + 2 \left| \sum_{\substack{\chi \in \mathbb{N}^n, \chi \neq 0 \\ \|\chi\| \leq p}} \sum_{\substack{\gamma \in \mathbb{N}^m \\ \|\gamma\| \leq p}} c_{\chi,\gamma^{(k_1)}} c_{\chi,\gamma^{(k_2)}} \right|} \end{bmatrix} \tag{6-53}$$

where $\gamma^{(k_1)}$ and $\gamma^{(k_2)}$ are two realizations of the index γ. These two equations provide two efficient and convenient approximations for the interval bounds. Since all the Chebyshev polynomials are not possible to achieve -1 or 1 at the same time, the estimations by (6-52) and (6-53) inevitably involve over-estimations. Proper algorithms, e.g. the scan method, can be used to better control the overestimation and obtain the bounds of the interval mean and standard variance as follows:

$$\mu(P(\boldsymbol{\xi}, \boldsymbol{\eta})) = \begin{bmatrix} \min \sum_{\gamma \in \mathbb{N}^m, \|\gamma\| \leq p} c_{0,\gamma} \Psi_\gamma(\boldsymbol{\eta}), \\ \max \sum_{\gamma \in \mathbb{N}^m, \|\gamma\| \leq p} c_{0,\gamma} \Psi_\gamma(\boldsymbol{\eta}) \end{bmatrix} \tag{6-54}$$

$$\sigma\left(P\left(\xi,\eta\right)\right) = \left[\frac{\sqrt{\min\limits_{\substack{\chi\in\mathbb{N}^{n},\chi\neq0 \\ \|\chi\|\leq p}}\left(\sum\limits_{\substack{\gamma\in\mathbb{N}^{m} \\ \|\gamma\|\leq p}}c_{\chi,\gamma}\Psi_{\gamma}\left(\eta\right)\right)^{2}},}{\sqrt{\max\limits_{\substack{\chi\in\mathbb{N}^{n},\chi\neq0 \\ \|\chi\|\leq p}}\left(\sum\limits_{\substack{\gamma\in\mathbb{N}^{m} \\ \|\gamma\|\leq p}}c_{\chi,\gamma}\Psi_{\gamma}\left(\eta\right)\right)^{2}}}\right] \tag{6-55}$$

6.6 Optimization Framework

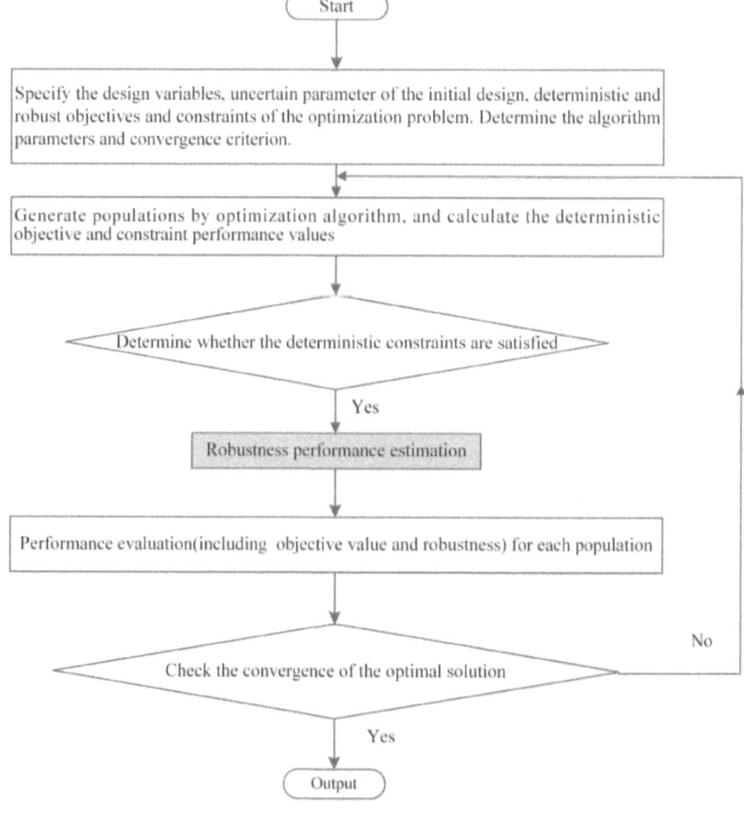

Figure 6-3 Flowchart of the general robust optimizer.

For robust optimization considering uncertainties, the calling of the performance calculation model is multiplied due to the requirement of robustness estimation for each solution in the iteration. In addition to the utilization of the effective uncertainty quantification methods as mentioned above, an effective strategy on reducing the potential solution number which requires the robustness assessment is emphasized in this research. The individuals in the population which cannot meet the deterministic constraints also cannot meet the robust constraints. If these solutions are eliminated firstly, the computation burden caused by the robustness measurement for all the individuals in the population can be relieved. Therefore, we developed the robust optimization framework based on the uncertainty analysis method and algorithms with the deterministic constraint filter. The robust optimization flowchart of the method is presented in Figure 6-3. It mainly includes the following steps.

Step 1:Specify the design variables, uncertain parameters of the initial design, deterministic and robust objectives and constraints of the problem. Determine the algorithm parameters and stop criterion.

Step 2: Generate a population for the evolutionary optimization algorithm, and calculate the deterministic objective and constraint.

Step 3: If the solution in the population cannot meet the requirements of the deterministic constraints, there is no need for this solution to do the reliability analysis with the uncertainty quantification method. Therefore, to accelerate the algorithm, the constraint performance of the population is evaluated in this step. Only if the constraints are satisfied, the reliability analysis is conducted in the following steps for the solution.

Step 4: Generate samples for the uncertainty analysis model building.

Step 5: Estimate the robustness of the solutions with the established uncertainty analysis model, and search the optimal solution in the population that meeting the robustness requirement

Step 6: Check whether the iteration condition is met of the current optimal solution, end the iteration until the requirement is met. Otherwise, go to step 2 to continue the iteration.

6.7 Design Example

To investigate the effectiveness of the proposed robust optimizers, a brushless DC wheel motor benchmark is selected as the design example [8]. The benchmark motor is designed for propelling a solar vehicle to meet the specifications of 20Nm torque at 721rpm while the maximal speed at no-load condition is 1442 rpm. The initial design is an outer rotor motor with surface mounted magnets, concentrated windings, 18 slots, and 12 poles. The motor topology is presented in Figure 6-4.

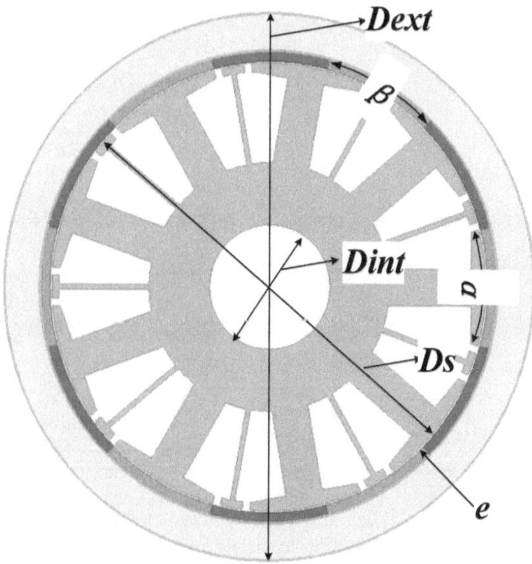

Figure 6-4 Topology of the brushless DC motor.

As a design benchmark, the example shows characteristics in explicit equations, scaled design variables, and objectives, etc. As well, it includes highly constrained multidisciplinary and multimodal features. The analytical design model consists of 78 nonlinear equations of the electromagnetic and thermal performances. The benchmark can be used for problems of both single and multiple objectives. When it is optimized as a single objective problem to achieve the highest efficiency under certain technical constraints, it can be defined as

$$\max \ \eta$$

$$s.t. \ \begin{cases} M_{tot} \leq 15kg, D_{int} \geq 76mm, I_{max} \geq 125A \\ D_{ext} \leq 340mm, T_{a} \leq 120°C, discr \geq 0 \end{cases} \qquad (6\text{-}56)$$

where M_{tot} is the total mass of the motor, D_{ext} the external diameter of the motor, D_{int} the inner diameter of the stator, I_{max} the maximum demagnetization current, T_{a} the temperature of the magnets, and *discr* the determinant used for the calculation of slot height.

Here we express the (6-56) with the following form

$$\max \ f_1 = \eta$$

$$s.t. \ \begin{cases} g_1 = 15 - M_{tot} \leq 0, g_2 = 76 - D_{int} \leq 0, \\ g_3 = 125 - I_{max} \leq 0, g_4 = D_{ext} - 340 \leq 0, \\ g_5 = T_{a} - 120 \leq 0, g_6 = discr \leq 0, \end{cases} \qquad (6\text{-}57)$$

While sharing the same constraints, for the multi-objective problem, another objective is to reduce the total mass which can be written as

$$obj. \ \begin{cases} \max \ f_1 = \eta \\ \min \ f_2 = M_{tot} \end{cases} \qquad (6\text{-}58)$$

Table 6-2 Design variables

Par.	Description	Unit	lower	upper
B_e	Maximum magnetic induction in the air gap	T	0.5	0.76
B_d	Average magnetic induction in the teeth	T	0.9	1.8
B_{cs}	Average magnetic induction in stator back iron	T	0.6	1.6
Ds	Stator outer diameter	mm	150	330
J	Current density	A/mm^2	2	5

In this design example, the five-design-parameter problem is selected for optimization. The design variables are listed in Table 6-2. The uncertain parameters are presented in

Table 6-3. The quantitive definition of the uncertain variable is set in the following sections according to the different types of uncertainties.

Table 6-3 Uncertain variables

Par.	Description	Unit
α	Width of stator tooth	deg.
e	Length of air gap	mm
β	Width of intermediate tooth	deg.
Ds	Stator outer diameter	mm
L_m	Length of the motor	mm

For the mono-objective problem, the DEA is selected as the optimization algorithm, while for the multi-objective case, the multi-objective GA is applied. For the robustness estimation, the methods introduced in this chapter and the conventional calculation methods are both adopted for comparison. They are PCE method, DR method and MCA for stochastic uncertainties, CI method and scan method for interval uncertainties, and PCCI method and SMCA method for hybrid uncertainties.

6.7.1 The Case with Stochastic Uncertainties

Assuming the uncertain parameters obey the normal distribution, their mean and standard deviation values are listed in Table 6-4. Since the stator outer diameter is also the design variable, the exact mean value is not listed and set as the solution of the population generated by the algorithm.

Table 6-4 Uncertain variables with stochastic uncertainty

Par.	Description	Unit	μ	σ
α	Width of the stator tooth	deg.	30	0.1
e	Length of air gap	mm	0.8	0.01
β	Width of intermediate tooth	deg.	6	0.1
Ds	Stator outer diameter	mm	-	0.1
L_m	Length of the motor	mm	45	0.1

The robust optimization model of the single objective problem is defined as

$$\begin{aligned}
\min : \ & F_1 = -\left[\mu_{f_1} + 6\sigma_{f_1}\right] \\
s.t. \quad & G_i = \mu_{g_i} + 6\sigma_{g_i} \leq 0 \\
& i = 1, \cdots, 6
\end{aligned} \tag{6-59}$$

where μ_{f_1} σ_{f_1} μ_{g_i} σ_{g_i} are the mean of objective, standard deviation of the objective, mean of the constraint, and standard deviation of the constraint, respectively.

While for another objective of the multi-objective model is expressed as

$$\min : \ F_2 = \left[\mu_{f_2} + 6\sigma_{f_2}\right] \tag{6-60}$$

6.7.1.1 Optimization Results of the Mono-Objective Problem

Table 6-5 lists the optimization solutions and objective function values of the benchmark obtained by the deterministic and robust optimizer based on MCA, PCE and DR method. The robust optimization approaches yield the same objective values and are very close to those obtained by the deterministic design optimization approaches which prove that the objective function is not of high sensitivity to the uncertain variables.

Table 6-5 Optimal design parameters and objective values

Par.	Unit	Deter.	MCA	PC	DR
B_d	T	1.8	1.8	1.8	1.8
B_e	T	0.650	0.655	0.665	0.665
B_{cs}	T	0.950	0.99	0.940	0.955
D_s	mm	202.5	201.0	199.5	199.5
J	A/mm^2	2.0415	2.0661	2.1579	2.1474
f_1		0.9531			
$-F_1$			0.9519	0.9519	0.519

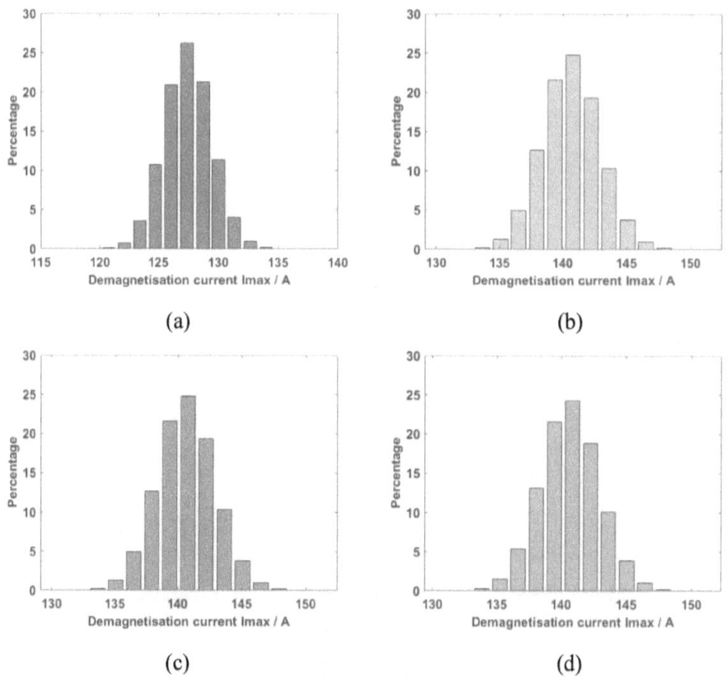

Figure 6-5 Demagnetization current distribution of the (a) deterministic solution, and robust solutions obtained by the optimizer based on (b) DR (c) MCA and (d) PCE.

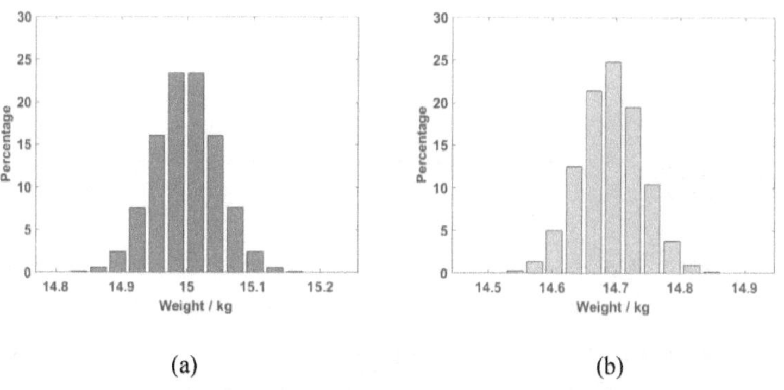

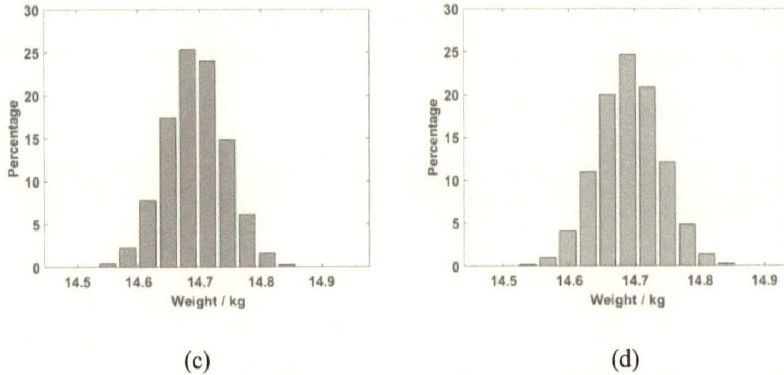

(c)   (d)

Figure 6-6 Total weight distribution of the (a) deterministic solution, and robust solutions obtained by the optimizer based on (b) DR method (c) MCA and (d) PCE.

Particularly, the demagnetization current that the permanent magnets can tolerate should be larger than 125 A, while the total weight is limited no more than 15 kg. Figure 6-5 and Figure 6-6 illustrate the distributions of the demagnetization current and weight for the deterministic and robust solutions obtained by the robust optimizer based on PC and DR. The results show that the possibility of the deterministic solution violates the constraints, while the robustness of solution with robust optimizers is well promised.

Table 6-6 Constraint violation condition of the solutions

Constraint	Deter.	MCA	PC	DR
G_1	0	1	1	1
G_2	1	1	1	1
G_3	0	1	1	1
G_4	1	1	1	1
G_5	1	1	1	1
G_6	1	1	1	1

Since the application of the DFSS, we utilized the direct sigma level to show if the robustness settings are met. The sigma level of optimal solutions obtained by different optimizers is calculated for verifying if the constraints are met by the MCA and the verification results are listed in Table 6-6, where 1 means the constraint is met and 0 means not. The verification results show that the robustness of the solutions obtained by the three robust optimizers is ensured. On the contrary, since there is no reliability assessor in the deterministic optimization, the robustness of the deterministic solution is not guaranteed.

Table 6-7 Mean value comparison of the robust solution obtained by the DR based optimizer

Performance	Unit	MCA	PCE	DR
η		0.9526	0.9526	0.9526
D_{int}	mm	79.96	79.96	79.96
D_{ext}	mm	238.13	238.13	238.13
discr		0.0247	0.0247	0.0247
I_{max}	A	140.628	140.628	140.628
Ta	°C	96.196	96.196	96.196
Mtot	kg	14.692	14.692	14.692

Table 6-8 Standard deviation comparison of the robust solution obtained by the DR based optimizer

Performance	Unit	MCA	PCE	DR
η		1.0592E-4	1.0588E-4	1.0585E-4
D_{int}	mm	6.5360E-4	6.5357E-4	6.5321E-4
D_{ext}	mm	2.0564E-4	2.0558E-4	2.0557E-4
discr		7.7444E-4	7.7419E-4	7.7419E-4
I_{max}	A	2.2513	2.2512	2.2509
Ta	°C	0.1255	0.1256	0.1255
Mtot	kg	0.0485	0.0485	0.0485

To illustrate the accuracy of the PC and DR method, the mean and standard deviation of the performances in the optimization model of the robust solution obtained by the DR based optimizer is selected and calculated by the three methods. Table 6-7 and Table 6-8 list the comparison results where only a finite number of negligible differences can be found, which proves the accuracy of the PC method and DR method in the moment information calculation.

From the perspective of efficiency, since the rate of convergence of MCA is of order $1/\sqrt{N}$, which means a high number of samples are required. In the MCA based optimizer, the sampling number is set as 10000, while the sample numbers for PC and DR are 42 and 16 which means a large number of samples are saved for the robustness estimation. In addition, the filtering in the algorithm to eliminate the potential solution required to do the robustness analysis increases the optimization speed further.

6.7.1.2 Optimization Results of the Multi-Objective Problem

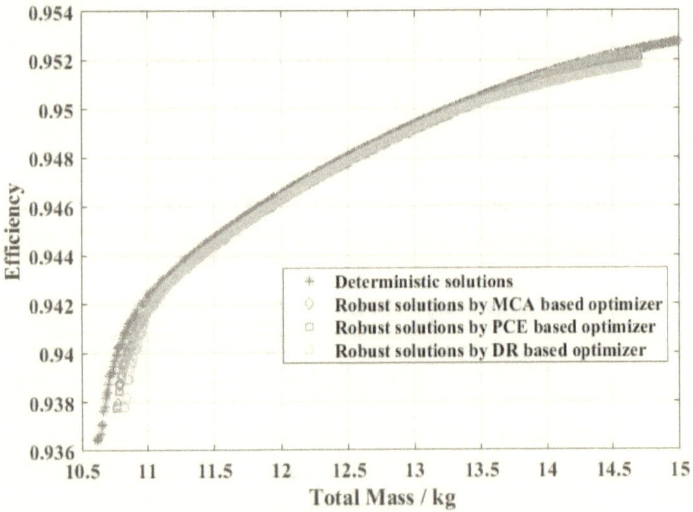

Figure 6-7 Pareto diagram of the deterministic and mean values of the objectives of the robust optimizations

Figure 6-7 illustrates the Pareto diagram of the deterministic and mean values of the objectives of the robust optimizations. As shown, robust solutions are more conservative than the deterministic solutions for achieving optimal efficiency and total weight. Meanwhile, the robust solutions achieved by the two optimizers are close to each other, confirming the accuracy of the proposed method.

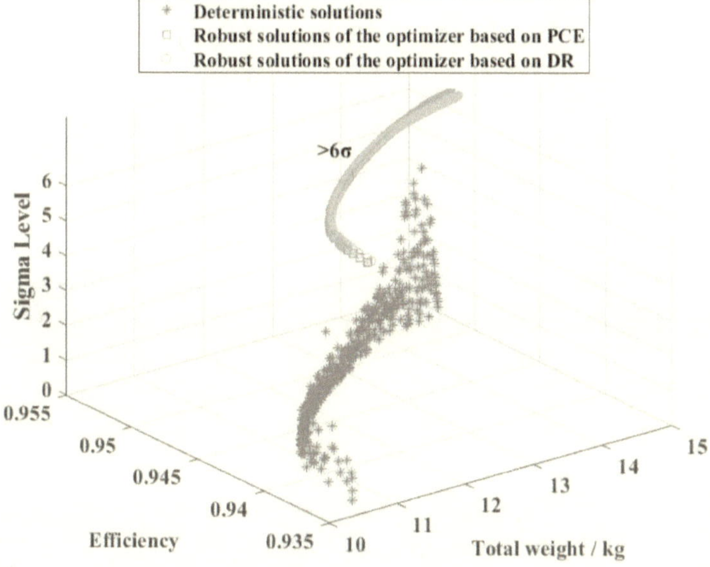

Figure 6-8 Sigma level of the solutions.

As further proof of the robustness of the solutions obtained by the PC and DR based robust optimizers, their sigma levels verified by the MCA are present in Figure 6-8. Particularly, the sigma level used is the minimum one among all constraints. In contrast to the deterministic solutions, the sigma levels of the robust solutions are higher than the pre-set value 6 which meet the robustness settings

To explore the performance variation interval with stochastic uncertainties, Figure 6-9 shows the error bar diagrams of the demagnetization current of the solutions. In each error bar, the nominal points are the mean value of the performance and the upper and lower bound are 6 times the standard deviation of the performance. As shown, various deterministic solutions violate the demagnetization current constraints, while the lower bounds of robust solutions are larger than 125 A.

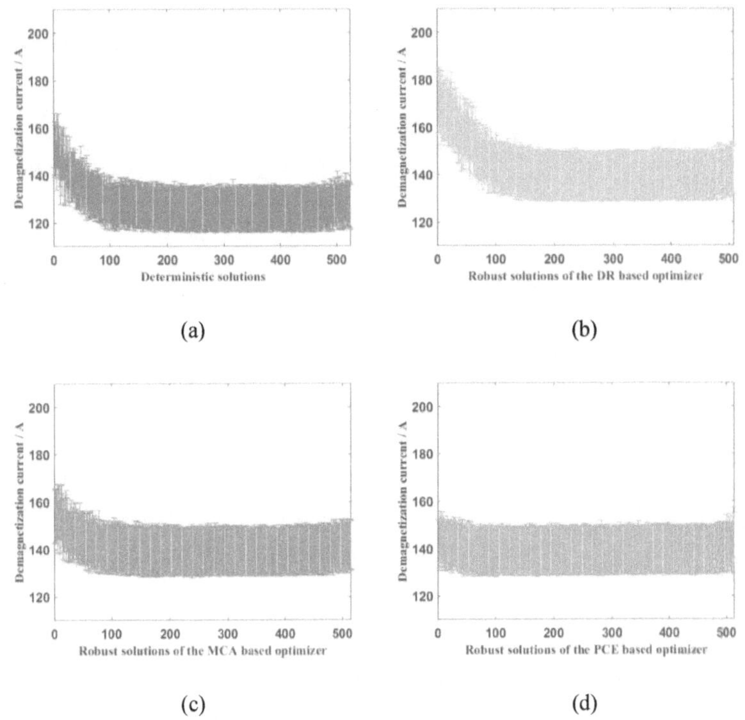

Figure 6-9 Demagnetization current distribution interval of the (a) deterministic solutions, and robust solutions obtained by the optimizer based on (b) DR (c) MCA and (d) PCE.

Table 6-9 Uncertain variables with interval uncertainty

Par.	Description	Unit	λ	δ
α	Width of stator tooth	deg.	30	0.6
e	Length of air gap	mm	0.8	0.06
β	Width of intermediate tooth	deg.	6	0.6
Ds	Stator outer diameter	mm	-	0.6
L_m	Length of the motor	mm	45	0.6

6.7.2 The Case with Interval Uncertainties

For the case with only the interval uncertainties, the nominal and bound values of the uncertain variables are listed in Table 6-9. The set of the bound values follows the rules of $\delta = 6\sigma$, where σ is the standard deviation of the uncertain parameters listed in Table 6-4.

Considering the worst-case scenario with interval uncertain variables, the robust optimization model of the single objective problem is defined as

$$\begin{aligned} \min: \ & F_1 = -\max(f_1) \\ s.t. \quad & G_i = \max(g_i) \le 0 \\ & i = 1,\cdots,6 \end{aligned} \tag{6-61}$$

While for another objective of the multi-objective model is expressed as

$$\min: F_2 = \max\left(f_2\right) \tag{6-62}$$

6.7.2.1 Optimization Results of the Mono-Objective Problem

Table 6-10 lists the optimization results obtained by the deterministic and robust optimizer based on the CI method and scan method. The two robust optimization approaches yield close objective values.

Table 6-10 Optimal design parameters and objective values

Par.	Unit	Deter.	Scan	CI
B_d	T	1.8	1.8	1.8
B_e	T	0.650	0.655	0.655
B_{cs}	T	0.950	0.99	0.975
D_s	mm	202.5	201.0	201.0
J	A/mm^2	2.0415	2.0661	2.0736
f_1		0.9531		
$-F_1$			0.9511	0.9510

To prove the accuracy of the CI method in estimation of the maximum and minimum values of the performances in the optimization model, their estimated values of the robust solutions obtained by both optimizers are verified by both CI and scan method.

Table 6-11 and Table 6-12 list the calculation results. The close minimum and maximum performance values validated by the CI method and scan method proved the accuracy of the CI method in the robust information estimation.

Table 6-11 Performance bound calculation of the robust solution of CI based optimizer

Performance	Unit	Maximum		Minimum	
		CI	Scan	CI	Scan
η		0.9530	0.9528	0.9510	0.9510
D_{int}	mm	94.07	92.76	79.45	79.72
D_{ext}	mm	241.07	241.07	236.74	237.02
$discr$		0.0364	0.0357	0.0190	0.0194
I_{max}	A	176.8056	176.8688	127.4292	132.07085
Ta	°C	98.0057	97.9329	94.8129	95.0485
$Mtot$	kg	14.9983	14.9964	13.9444	14.0198

Table 6-12 Performance bound calculation of the robust solution of scan based optimizer

Performance	Unit	Maximum		Minimum	
		CI	Scan	CI	Scan
η		0.9531	0.9530	0.9511	0.9511
D_{int}	mm	92.92	91.57	78.06	78.33
D_{ext}	mm	240.65	240.65	236.39	236.66
$discr$		0.0360	0.0354	0.0185	0.0190
I_{max}	A	169.8452	169.9022	122.9301	127.3059
Ta	°C	98.0764	98.0034	94.8421	95.0834
$Mtot$	kg	14.9999	14.9997	13.9617	14.0537

Table 6-13 Constraint violation condition of the solutions

Constraint	Deter.	Scan	CI
G_1	0	1	1
G_2	0	1	1
G_3	0	1	1
G_4	1	1	1
G_5	1	1	1
G_6	1	1	1

Table 6-13 lists the constraint violation condition of the solutions verified by the scan method considering the worst-case scenario. The verification results show that the robustness of the solutions obtained by the robust optimizers is ensured.

From the perspective of efficiency, in addition to the filtering setting of the algorithm, the number of sample points for the robustness estimation of the CI method is 42. The accuracy of the scan method is up to the sampling number. In the example. 10 symmetrical points in the variation interval for each parameter are sampled, which means 100000 samples are required. For many cases of complex models, the huge sample number required by the scan method is unacceptable. The comparison of the sampling number proves the merits of the CI based optimizer.

6.7.2.2 Optimization Results of the Multi-Objective Problem

Figure 6-10 illustrates the Pareto diagram of the deterministic and robust optimizations achieved by the two robust optimizers. As shown, robust solutions are more conservative than deterministic solutions for achieving optimal efficiency and total weight. Meanwhile, the robust solutions achieved by the two optimizers are close to each other, confirming the accuracy of the proposed method.

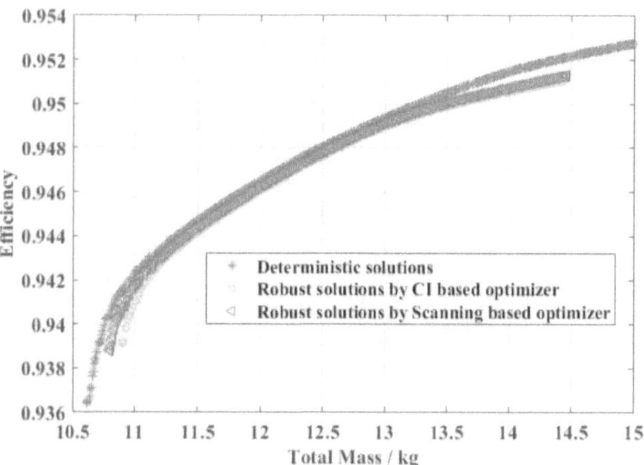

Figure 6-10 Pareto diagram of the deterministic optimization and nominal values of the robust optimization.

As further proof of the robustness of the solutions obtained by the robust optimizer based on the CI method, their feasibility is verified by the scan method and compared with the deterministic solutions. The feasibility represents whether the constraints are met under the interval uncertainties. As shown in Figure 6-11, in contrast to the deterministic solutions, the feasibility of the solutions achieved by the proposed robust optimizer is well promised.

Figure 6-12 presents the error bar diagram comparison of the robust solutions and deterministic solutions. The nominal values of the performance are achieved by considering the nominal values of interval uncertain parameters only while the lower bound and upper bound are calculated by the scan method. The violating of the constraints can be observed by deterministic optimal solutions.

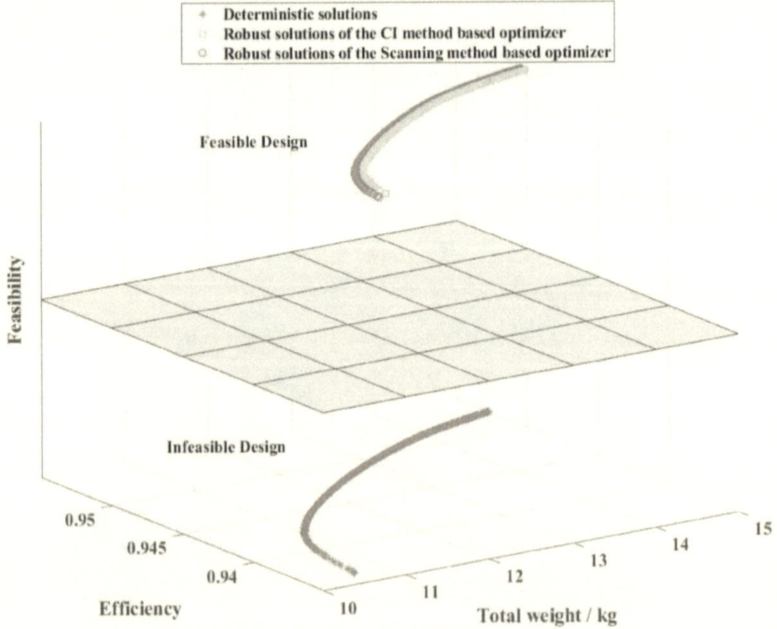

Figure 6-11 Feasibility of the solutions.

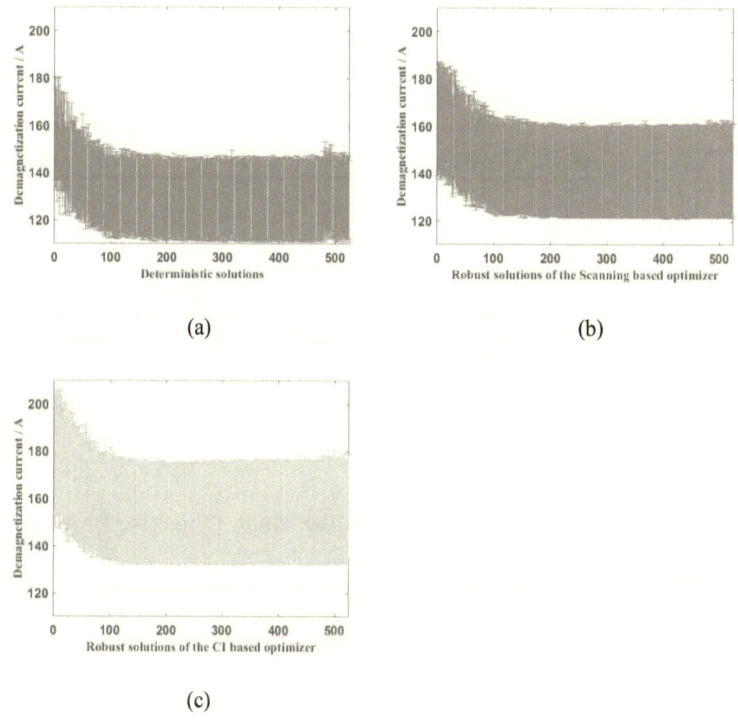

Figure 6-12 Demagnetization current distribution interval of the (a) deterministic solutions, and robust solutions obtained by the optimizer based on (b) scan method, and (c) CI method.

6.7.3 The Case with Hybrid Uncertainties

For the case of the robust optimization problem with hybrid uncertainties, the uncertain variables are divided into two groups with interval and stochastic uncertainties respectively as presented in Table 6-14 and Table 6-15. In order to prove the feasibility of the proposed optimizer, the deterministic optimization and robust optimization based on the SMCA are conducted for comparison.

Table 6-14 Uncertain variables with interval uncertainty

Par.	Description	Unit	λ	δ
α	Width of stator tooth	deg.	30	0.6
e	Length of air gap	mm	0.8	0.06
β	Width of intermediate tooth	deg.	6	0.6

Table 6-15 Uncertain variables with stochastic uncertainty

Par.	Description	Unit	μ	σ
Ds	Stator outer diameter	mm	-	0.1
Lm	Length of the motor	mm	45	0.1

6.7.3.1 Optimization Results of the Mono-Objective Problem

Table 6-16 Optimal design parameters and objective values

Par.	Unit	Deter.	PCCI	SMCA
B_d	T	1.8	1.8	1.8
B_e	T	0.650	0.67	0.67
B_{cs}	T	0.950	1.035	1.135
D_s	mm	202.5	199.5	201.0
J	A/mm^2	2.0415	2.1951	2.1858
f_1		0.9531		
$-F_1$			0.9513	0.9512

Table 6-16 lists the optimization results of the benchmark obtained by the deterministic and robust optimization methods with PCCI and SMCA. The two robust optimization approaches yield the same objective values, which are very close to those obtained by the deterministic design optimization approaches.

To verify the accuracy of the PCCI method in the robustness estimation, the relevant performance of the robust solution obtained the PCCI based optimizer verified by SMCA. Table 6-17 and Table 6-18 present the calculation results for the robustness estimation, where MU, ML, SU, and SL are upper-bound mean, lower-bound mean, upper-bound standard deviation, and lower-bound standard deviation of the performance respectively. The comparison results demonstrate the accuracy of the PCCI method. The constraint violation status of the solutions is listed in Table 6-19. The verification results show that the robustness of the solutions obtained by the optimizers is ensured.

Table 6-17 Calculation results of the SMCA

Performance	Unit	MU	ML	SU	SL
η		0.9529	0.9513	2.008E-5	1.54E-5
D_{int}	mm	89.10	80.61	3.6820E-1	2.927E-1
D_{ext}	mm	239.95	237.15	1.184E-1	1.132E-1
$discr$		0.0328	0.0209	3.324E-4	3.123E-4
I_{max}	A	167.3182	127.3147	0.369	0.2755
Ta	°C	97.3926	95.5379	0.0786	0.0671
$Mtot$	kg	14.8892	14.1379	0.0172	0.0157

Table 6-18 Calculation results of the PCCI method

Performance	Unit	MU	ML	SU	SL
η		0.9529	0.9513	1.972E-5	1.561E-5
D_{int}	mm	89.10	80.61	3.601E-1	2.972E-1
D_{ext}	mm	239.95	237.15	1.629E-1	1.1569E-1
$discr$		0.0328	0.0209	3.260E-4	3.187E-4
I_{max}	A	167.295	127.346	0.365	0.279
Ta	°C	97.3897	95.5406	0.0776	0.0681
$Mtot$	kg	14.8890	14.1384	0.0171	0.0159

Table 6-19 Constraint violation condition of the solutions

Constraint	Deter.	SMCA	PCCI
G_1	0	1	1
G_2	0	1	1
G_3	0	1	1
G_4	1	1	1
G_5	1	1	1
G_6	1	1	1

From the perspective of efficiency, for the interval and stochastic parameters, the number of the scan and Monte Carlo sampling points is 1,000 respectively, which means that the number of total sampling points is 10^6 while only 60 (the coefficient number) samples are required for the second-order PCCI modeling. Please note, as we indicated above, twice the coefficient number is suggested for numerical stability in other problems.

6.7.3.2 Optimization Results of the Multi-objective Problem

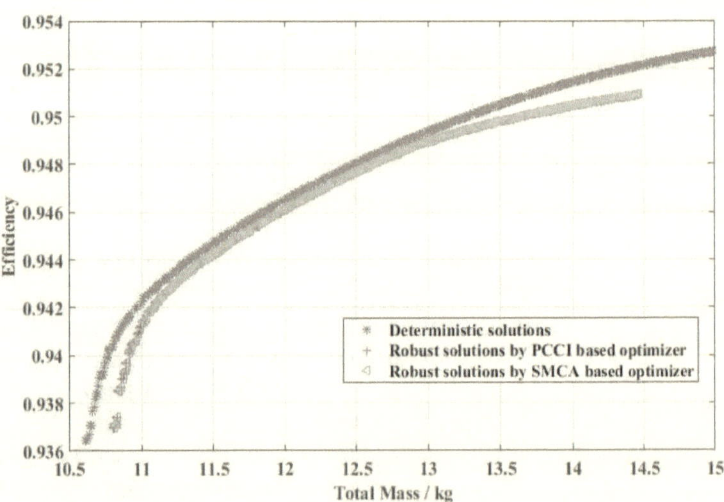

Figure 6-13 Pareto diagram of the deterministic and robust optimizations.

Figure 6-13 illustrates the Pareto diagram of the deterministic and robust optimizations. As shown, robust solutions are more conservative than deterministic solutions for achieving optimal efficiency and total weight. Meanwhile, the robust solutions achieved by the two optimizers are close to each other, which confirms the accuracy of the proposed method. As further proof of the robustness of the solutions obtained by the PCCI based optimizer, their feasibility verified by the SMCA is present in Figure 6-14. In contrast to the deterministic solutions, the feasibility of the solutions achieved by the proposed robust optimizer is well promised.

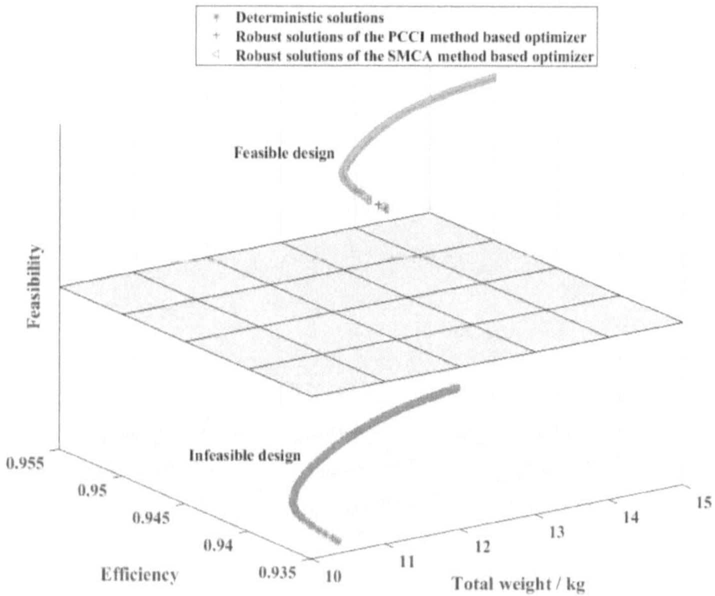

Figure 6-14 Feasibility of the solutions.

Figure 6-15 shows the error bar diagrams of the demagnetization current of the solutions, the length of the bar is defined as $\mu_{ub}(I_{max}) - \mu_{lb}(I_{max}) + 2n_{f}\sigma_{ub}(I_{max})$. As shown, various deterministic solutions violate the demagnetization current constraints, while the lower bounds of robust solutions are larger than 125 A.

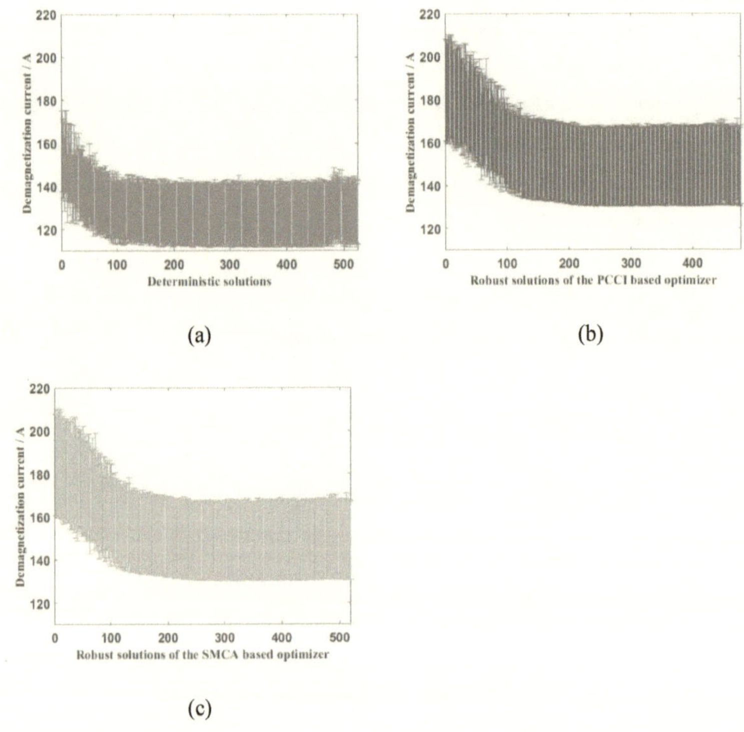

Figure 6-15 Demagnetization current distribution interval of the (a) deterministic solutions, and robust solutions obtained by the optimizer based on (b) SMCA method, and (c) PCCI method.

6.8 Conclusions

For efficient robust optimization of electrical machine design considering stochastic, interval, and hybrid uncertainties, this research introduces the uncertainty quantification methods into a uniform robustness optimization model. Compared with classical methods such as MCA, scan, and SMCA, these methods show comparable accuracy and much higher efficiency. Then a general framework with efficient filtering for increasing the optimization speed further. The numerical optimization results and comparisons of the design example prove the superiority of the proposed robust optimizers.

Chapter 7 Conclusions and Future Works

7.1 Conclusions

This Book focused on the development of design and optimization techniques for electrical machines. A comprehensive literature survey is presented in Chapter 2. The design optimization modeling, basic procedure, and effective techniques and strategies are reviewed. Various case studies are presented for the verification of the effectiveness of the presented techniques. As a specific design example, the development of a new dual rotor axial flux motor with grain-oriented silicon steel core material for the in-wheel drive application is presented in Chapter 3 and Chapter 4, which includes the application analysis, material selection, electromagnetic design, prototyping, test, and optimization.

According to the state-of-art investigation on the design optimization of electrical machines, challenges and proposals on topology optimization, optimization considering manufacturing, and application-oriented system-level design optimization for electrical machines are also introduced in Chapter 2. This study focuses on the investigation of the former two optimization methods.

Chapter 5 presents the work on topology optimization for ferromagnetic materials of electrical machines. Compared with the parametric size optimization and shape optimization, topology optimization shows the highest freedom. A general topology optimization method for the ferromagnetic components in PM electrical machines based on the density method is developed. The optimization results of two examples prove the effectiveness of the proposed method.

Chapter 6 shows the efficient general optimization method development considering stochastic, interval, and hybrid uncertainties exist in manufacturing. Even with different uncertainties, the robustness is modeled in a unified form. With the introduced uncertainty quantification methods for different uncertainties and algorithms with the deterministic constraint filter, the proposed new robust optimizers show high feasibility for the robust optimization of electrical machines.

7.2 Future Works

The future work will be conducted on the following aspects.

(1) Performance enhancement of the dual rotor axial flux in-wheel motor by more comprehensive design such as the water-cooling structure, multidisciplinary analysis, manufacturing method, experiment and optimization schemes are to be proposed.

(2) Investigation of the topology optimization considering more performances such as the loss, efficiency, and multi-materials, multidisciplinary, three-dimensional structure with more topology optimization methods such as level set method.

(3) Development of robust optimization method considering high dimensional uncertainties, effective algorithms, data-driven uncertainty quantification methods.

(4) Investigation of the application-oriented system-level design optimization techniques considering multidisciplinary performances.